T0342120

Data Centre Essentials

Data Centre Essentials

Design, Construction, and Operation of Data Centres
for the Non-expert

Vincent Fogarty

Fellow of the Royal Institute of Charted Surveyors
Incorporated Engineer, Engineering Council of the UK

Sophia Flucker

Chartered Engineer, Engineering Council of the UK
Member of the Institution of Mechanical Engineers

Registered Offices
John Wiley & Sons, Inc., 111 River Street, Hoboken, NJ 07030, USA
John Wiley & Sons Ltd., The Atrium, Southern Gate, Chichester, West Sussex, PO19 8SQ, UK

For details of our global editorial offices, customer services, and more information about Wiley products visit us at www.wiley.com.

Library of Congress Cataloging-in-Publication Data
Names: Fogarty, Vincent, author. | Flucker, Sophia, author.
Title: Data centre essentials : design, construction, and operation of data
   centres for the non-expert / Vincent Fogarty and Sophia Flucker.
Description: Hoboken, NJ : Wiley-Blackwell, 2023. | Includes index.
Identifiers: LCCN 2023007219 (print) | LCCN 2023007220 (ebook) | ISBN
   9781119898818 (hardback) | ISBN 9781119898825 (adobe pdf) | ISBN
   9781119898832 (epub)
Subjects: LCSH: Data centers.
Classification: LCC TJ163.5.D38 F64 2023 (print) | LCC TJ163.5.D38
   (ebook) | DDC 005.74–dc23/eng/20230224
LC record available at https://lccn.loc.gov/2023007219
LC ebook record available at https://lccn.loc.gov/2023007220

Cover Image: Wiley
Cover Design: © Yuichiro Chino/Getty Images

Set in 9.5/12.5pt STIXTwoText by Straive, Pondicherry, India
Printed and bound by CPI Group (UK) Ltd, Croydon, CR0 4YY

C9781119898818_141124

Contents

Acronyms & Symbols

ACoP L8	Approved Code of Practice L8
Ag	Silver
AHU	Air Handling Unit
AI	Artificial Intelligence
AIA	American Institute of Architects
AP	Authorised Person
ARP	Alarm Response Procedure
ASHRAE	American Society of Heating, Refrigeration and Air Conditioning Engineers
ATP	Authorisation to Proceed
ATS	Automatic Transfer Switch
BIM	Building Information Modelling
BMS	Building Monitoring System or Building Management System
BREEAM	Building Research Establishment Energy Assessment Methodology
BSI	British Standards Institute
BSRIA	Building Services Research and Information Association
CAPEX	Capital Expenditure
CBEMA	Computer and Business Equipment Manufacturers Association
CCU	Close Control Unit
CDM	Construction Design and Management
CDP	Contractor Design Portion
CFD	Computational Fluid Dynamics
CLS	Cable Landing Station
CPU	Central Processing Unit
CRAC	Computer Room Air Conditioning
CRAH	Computer Room Air Handling
CSP	Cloud Service Provider
Cu	Copper
DCF	Discounted Cash Flow

DCIM	Data Centre Infrastructure Management
Delta T	Difference in Temperature
DNS	Domain Name System
DP	Dew Point
DX	Direct Expansion
EBC	Energy in Buildings and Communities Programme
EIA	Environmental Impact Assessment
EIS	Environmental Impact Statement
EMI	Electromagnetic Field Interference
EN50600	European Standard 50600: Information Technology. Data Centre Facilities and Infrastructures
EOP	Emergency Operating Procedure
EPC	Engineer, Procure, Construct
ERs	Employer Requirements
ESG	Environmental, Social, and Governance
ETS	Emissions Trading Scheme
EU	European Union
FAC-1	Framework Alliance Contract
FAT	Factory Acceptance Testing
FLAP	Frankfurt, London, Amsterdam, and Paris
FIDIC	International Federation of Consulting Engineers
FM	Facilities Management
FMECA	Failure Modes, Effects, and Criticality Analysis
FPT	Functional Performance Testing
FWT	Factory Witness Testing
GIM	Gross Income Multipliers
GPP	Green Public Procurement
GPU	Graphics Processing Unit
GRI	Global Reporting Initiative
HFC	Hydrofluorocarbon
HSE	Health and Safety Executive
HSSD	High Sensitivity Smoke Detection
HV	High Voltage
HVO	Hydrotreated Vegetable Oil
IC	Integrated Circuit
ID	Identity Documentation
IEA	International Energy Agency
IEC	Indirect Evaporative Cooler
IFC	Issued For Construction
IoT	Internet of Things
IP	Internet Protocol

IRC	In-Row Cooler
IRL	Issues Resolution Log
ISO	International Organization for Standardization
IST	Integrated Systems Testing
IT	Information Technology
ITIC	Information Technology Industry Council
ITUE	IT Power Usage Effectiveness
KPI	Key Performance Indicator
kW	Kilowatt
LCA	Life Cycle Assessment
LEED	Leadership in Energy and Environmental Design
LOTO	Lock Out Tag Out
LV	Low Voltage
M&E	Mechanical and Electrical
MMR	Meet Me Room
MOP	Maintenance Operating Procedure
MPOP	Main Points of Entry
MSL	Managed Service Provider
MTS	Manual Transfer Switch
MTTR	Mean Time To Repair
MW	Megawatt
MWh	Megawatt Hour
N	Need
NABERS	National Australian Built Environment Rating System
NDA	Non-Disclosure Agreement
NEC	New Engineering Contract
NIM	Net Income Multipliers
NIS	Natural Impact Statement
NNN	Triple Net Lease
NOC	Network Operations Centre
NOI	Net Operating Income
NOx	Nitrogen Oxides
O&M	Operating and Maintenance
OFCI	Owner Furnished, Contractor Installed
OPEX	Operational Expenditure
OPR	Owner's Project Requirements
PC	Practical Completion
PCB	Printed Circuit Board
PDU	Power Distribution Unit
PoP	Point of Presence
PPA	Power Purchase Agreements

PPC2000	Project Partnering Contracts, 2000
PPE	Personal Protective Equipment
PPM	Planned Preventative Maintenance
PTW	Permit to Work
PUE	Power Usage Effectiveness
RAMS	Risk Assessment and Method Statement
RCA	Root Cause Analysis
RCM	Reliability-Centred Maintenance
RDC	Rear Door Cooler
REITs	Real Estate Investment Trusts
RFI	Request for Information
RFS	Ready for Service
RH	Relative Humidity
RIBA	Royal Institute of British Architects
RICS	Royal Institute of Chartered Surveyors
ROM	Rough Order of Magnitude
RPO	Recovery Point Objective
RTO	Recovery Time Objective
SAP	Senior Authorised Person
SBTi	Science-Based Targets Initiative
SDG	Sustainable Development Goal
SDIA	Sustainable Digital Infrastructure Alliance
SLA	Service-Level Agreement
SLTE	Submarine Line Termination Equipment
SOP	Standard Operating Procedure
SPOF	Single Point of Failure
SPV	Special Purpose Vehicle-Type Company
STS or S/S	Static Transfer Switch
TC9.9	Technical Committee 9.9
TCO	Total Cost of Ownership
TTRS	Time to Restore Service
TUE	Total Power Usage Effectiveness
UK	United Kingdom
UN	United Nations
UPS	Uninterruptible Power Supply
VPN	Virtual Private Network
WBCSD	World Business Council for Sustainable Development
XaaS	Anything as a Service

Foreword by Tom Glover

We live in exceptional times. The world's technological evolution is springboarding civilisation beyond nature's Darwinism limits. While the world deals with world issues, the world's ultrarich are heading for the stars and the next frontier – wherever this is, rest assured there will be a data centre nearby.

Keeping up with all the noise around us and ongoing change is a full-time job, which is why, if nothing else, we should all once in a while pause and remind ourselves that in order to manage the changes around us, we first need to understand the fundamental, underpinning enablers to those very changes.

The growth of the digital world has been rapid, and the change it has brought about is immense. Do you recall the first Sinclair ZX81 with 1K of RAM; the birth of email, modems, newsgroups, the internet, Amazon, Google, Microsoft, Apple, Facebook, and Salesforce all coming into being; and watching companies acquired, falling by the way, and morphing into different visions of themselves?

Every year we consume more digital experiences than the last, create more data, build more applications, and solve more problems through technology. When we lose our phone under the sofa, or our computer crashes, we feel lost, disconnected from humanity's modern hive. In many ways, in the way Tim Marshall talks about 'Prisoners of Geography,' are we not also 'Prisoners of Technology,' albeit technology's borders are global?

Throughout the journey of this digital evolution, one enabler has remained constant, adapting and improving, but always there 'on' in the background making our digital world so.

The Data Centre.

Without data centres, mobile phones do not work, communications revert to tapped-out messages and pigeon carriers, and global pandemics would have been tenfold worse – our world coming to an even harder 'STOP' than it did.

Behind everything we do at the heart of the digital world is the data centre. A purpose-built facility, housing the vast compute and connectivity capabilities needed to run our world. And yet how much do you really know about this silent global enabler?

It was in the 1990s, two years into my technology career that I first entered a data centre at BT Martlesham, the platform for science fiction, highly secure, a whirl of activity, flashing lights, and a background noise and smell that anyone who has visited a data centre will know. At the time I did not fully understand the importance of these non-descript buildings of infrastructure. Today I am a little more versed, still learning, and in awe of their silent importance in the world around us.

In 2009, I joined Interxion (now Digital Realty) working for one of our industries' veterans, David Ruberg, and my immersion into the world and inner workings of data centre began in earnest. I wish this book had been written then as it provides a roadmap for anyone interested in data centres, from employees, investors, developers, and operators to the mighty cloud providers. This book unpacks and simplifies the data centre into the constituent parts of a data centre's life.

I have had the pleasure to work with Vincent on several data centre projects; he has been an ambassador for the sector for well over two decades and exudes a humble passion for it. Vincent could have written each chapter in this book but chose to engage with the best within our industry who together have written chapters on their expert experiences, which when combined present a cohesive best-in-class insight into data centres.

Experts assembled, our own DC Titans so to speak, this data centre handbook of handbooks takes rudimentary topics that form the foundations of our digital world and brings them to life for the novice and expert alike. This book is for the reader who wants to understand how these inconspicuous buildings come into being and are operated, maintained, and protected – from bricks and mortar to penmanship.

If you are interested in getting data centre zoning or planning, securing power to a site, wanting to design and build a data centre, running the operating company, maintaining the data centre, wishing to unpick the legal complexities of a lease agreement, or simply understanding what is behind all the change around them in today's digital world, then this is the book.

We live in exceptional times; data centres are the exceptional infrastructures behind our world today, and this book will help you understand them.

Preface

The origins of this book commenced from an icy place in the subarctic. In the winter of 2015 and the spring of 2016, Sophia and I had to work on the same Facebook data centre project in Luleå, Sweden. Our contribution to the project was entirely different. I was working on commercial matters whilst Sophia was engaged in providing mechanical engineering support for design, commissioning, and operational cooling challenges on the previous and current phases of the project.

In this challenging environment where temperatures reached minus 34 °C, the project teams started early and finished late into the night and then, on many occasions, off to the restaurant and sometimes the only pub in Luleå. Most of those who worked on the project have never forgotten the harsh condition of living there; some meet up regularly. The nuances of living in Swedish houses with frozen locks, plug-in motor cars that prevent overnight freezing, and other constraints are not easily forgotten by those who were there.

As and when people experience all extreme challenges, people bonding and comradery follow. Sophia was more of the fly-in and -out expert, usually returning home at the weekend; however, she was very much part of the engineering team and a deeply respected solutions proposer. You do not need to spend much time with Sophia to recognise her unique talent for making complex engineering principles easy to digest by non-experts.

Following on from Luleå, we occasionally crossed paths at various industry events and had the old conversation here and there. Whilst my origins are more routed in engineering, I am more of a generalist in the data centre space, focusing on commercial matters. Whilst Sophia works as a technical expert in engineering matters, from those diverse positions appears a recognised void – the need for something for the non-expert who needs a holistic overview of the complete life cycle of a data centre.

When I asked Sophia if she would partner with me on this endeavour, her response was almost immediate, and we agreed on clear objectives. We soon recognised we needed an IT expert. So Max Schulze joined, covering the known and the unknowns about IT. The tentacles of data centres impact many legal transactions, and sometimes disputes emerge; those all need specialist legal services, and, in that regard, it was with great relief that Andrew McMillian joined our journey to set the legal background.

Tom Glover, who has held many senior positions at the most prominent global data centre practices, kindly agreed to provide the Foreword.

The key deliverable stemmed from the recognition that there was a need for those that may become involved in data centres but were on the periphery of that industry to understand the basics of data centres. Therefore, the book's vision is to provide knowledge to those who intend to provide professional service and other deliverables to the data centre industry but do not have a technical background or sector experience. The intention is that this book will give the essentials and address the entire domain context of the data centre life cycle from initial concept and investment appraisal to operational use.

The intention is to get a lot covered on a broad domain, but because data centres can be considered the 'brains' of the internet and are central to the digital economy, we encompass the wide margins of business needs and a whole range of dependable interfaces.

This book will outline the technical landscape and help identify the unknown unknowns, to help the reader understand where they will need to engage with professional data centre expert specialist services. In that context, it provides a high-level overview of critical considerations but is not a technical deep dive into complex matters of complex design, construction, commissioning, and operation.

Our objective is to fill the knowledge void for professionals like lawyers, financial and insurance advisors, surveyors, engineers, architects, and other professionals, who do not necessarily have a deep domain experience in data centres. It ought to assist companies that are equity investors, financial services firms, lessors of data centre space, insurance companies, facility management companies, data centre supply chain businesses, educators, and all those working within or on the periphery of the data centre industry. The broad domain of this book is intended to assist non-data centre professionals in gaining awareness of critical concepts and terminology, complement their core skills when drafting documents or making business decisions, and help them ask more informed questions when working on data centre projects.

Data centre solutions evolve so quickly that there is no perfect time to write about them. However, some of the fundamentals remain the same – the need for resilient power, cooling, and connectivity. Our key objective was to weave the narratives of data centre needs, site selection, engineering solutions, and legal context.

This book recognises that climate change is the defining challenge of our time, and the rehabilitation of the image of data centres as environmental bad boys to a sector that is keen to make steady inroads on sustainability continues. As we peek around the corner of the future, we recognise that regulation may need to be enacted and applied across borders consistently and transparently. What the future holds will no doubt give the need for future additions and revisions.

We drafted this book against extensive transitions following COVID impacts that accelerated home working, a platter of new internet needs, and the intensification of sustainability and global warming debates. For a book whose origins started in the freeze of the Nordics, data centres are now the hottest topic of our planet's engineered solutions.

Acknowledgements

This book benefits from being reviewed by a number of colleagues and collaborators. It would not have been possible without the support of a number of people.

Sophia Flucker
Thanks; Beth Whitehead, Marc Beattie, Niklas Sundberg, Robert Tozer, Steve Avis, and the whole team at OI and MiCiM who are a pleasure to work with and have taught me so much. And, of course, my nearest and dearest behind the scenes (who know who they are).

Vincent Fogarty
Before setting off on this journey, I asked John Mullen, the director of Driver Group Plc., himself a published co-author of Wiley for 'Evaluating Contract Claims,' for his initial insights on the writing process that have proved invaluable. Mark Wheeler, the CEO of Driver Group Plc., was motivational and supportive of my efforts of getting the book across the line.

Selina Soong, a consultant of Currie & Brown, who shared insightful knowledge gained from acting as the auditor of global data centre leases was gratefully received.

Lee Smith provided my graphical representation and was a pleasure to work with.

Max Schulze
I would like to dedicate this work to both my daughters June and Violet, who are always incredibly patient with me when I am caught up in my own thoughts. And, of course, none of the work would be possible without the discussions and dialogue within our NGO and the team of the Sustainable Digital Infrastructure Alliance (SDIA) and the wonderful community that over the years has informed many of the insights I am writing about in this book. A special thanks goes to Daan Terpstra, who is the CEO of the SDIA, for giving me the freedom to pursue projects like this book.

Many of the ideas I wrote about are the results of discussions and conversations with many people in the industry and beyond. Yet there is one person who has been a continuous source of inspiration: Jurg van Vliet. I am glad to be able to call him a friend.

Similarly, I would like to thank John Laban, who has been a strong supporter of the SDIA and who put me on the track of sustainable digital infrastructure, to begin with. I will never forget the first very long-hour encounter at a conference, with his pen and phone as the whiteboard. Another supporter and one of the first people working with me on the topic of sustainability in digital infrastructure is Mohan Gandhi; many of the insights from this book come from long debates we have had together during his time at the SDIA.

I would also like to thank the people of Perugia, Italy, for protecting the beautiful nature, which was the backdrop to most of my writing for this book.

Andrew McMillan
Acknowledges the Pinsent Masons LLP team and contributions from the following individuals spanning diverse legal practice areas: Becca Aspinwall, Anne-Marie Friel, Rob Morson, Mark Marfe, Angelique Bret, Paul Williams, and Tadeusz Gielas.

About the Authors

Vincent Fogarty received his MSc (King's College) in Construction Law & Dispute Resolution, BSc (Hons.) in Cost Management in Building Engineering Services. Diploma in Engineering Services, FRICS, FSCSI, MCIArb, ACIBSE, MIEI, I.Eng.

Vincent is a dual-qualified engineer and quantity surveyor and started his data centre journey as a mechanical and electrical consultant for the Bank of Ireland data centre in Dublin, which housed reel-to-reel data storage for some of the first cash machines. Many years later, Vincent ventured to Luleå in North Sweden as commercial manager on Facebook's first data centre built outside the United States. Since then, Vincent has provided commercial advice and dispute resolution services on various commercial matters on many data centre projects in many jurisdictions. Vincent is also a founding partner of Project Art, a data centre site currently under a planning application process in Ireland.

Vincent has over 38 years of combined quantity surveying and mechanical and electrical engineering experience within the construction industry. He initially trained as a mechanical and electrical engineer with a firm of specialist consultants and later joined consultants and contractors working on the whole project life cycle from inception to completion and then handover and operation. Lawyers have appointed Vincent as an expert in complex mechanical and electrical quantum matters concerning commercial data centre disputes. He has also given a quantum opinion on operational costs and loss of revenue in energy generation.

He is a fellow of the RICS since 2014 and maintains membership in the Institute of Engineers as an incorporated engineer. He recently became a member of the Chartered Institute of Arbitrators, having an MSc in Construction Law and Dispute Resolution from King's College, London.

Sophia Flucker is a Chartered Engineer with a background in mechanical engineering who has worked with data centres since 2004. She has worked as an engineering consultant on a variety of projects in several countries, including delivery of design, commissioning, training, and risk and energy assessments. This includes working alongside operations teams to help clients manage their critical environments and enjoy collaborating with others to optimise system performance. Her experience includes developing low-energy data centre cooling solutions, creating analytical and reporting tools, and leading teams and business development. Sophia's contributions to the digital infrastructure sector were recognised in 2020 when she received an Infrastructure Masons IM100 Award. She is passionate about sustainability and participates in various industry groups and has published and presented several technical papers on data centre energy efficiency and environmental impact. Sophia is the Managing Director of Operational Intelligence (OI), a company she joined at its inception in 2011, which delivers services to help optimise data centre reliability and energy performance throughout the project cycle. OI's approach focuses on the human element and improving knowledge sharing for better outcomes. In 2022, OI merged with MiCiM, specialists in mission-critical management, and Sophia joined their board as Technical Director.

Max Schulze is the Founder of the Sustainable Digital Infrastructure Alliance (SDIA). With a background as a software engineer and a cloud expert, Max brings his experience in measuring the digital economy's footprint and advancing the SDIA's roadmap towards making it sustainable and future-proof. Throughout his career, he has dedicated himself to the well-being of people and the planet as he is committed to creating a positive future for the next generation.

Andrew McMillan is a partner at an international law firm, Pinsent Masons LLP. Within Pinsent Masons, Andrew heads up the Technology & Digital Markets and Data Centre practices. He specialises in corporate transactions within the technology, media, and telecommunications space and advises trade players, investment banks, and private equity houses on transformative mergers and acquisitions, joint ventures, and high-value commercial partnering arrangements. He has been recommended by various legal directories, including Legal 500, Chambers, Who's Who Legal (Telecoms and Media), and Super Lawyers, both for his sector expertise and execution capability.

For several years, Andrew chaired an artist-led charity that was established to provoke and inform the cultural response to climate change, bringing together artists and scientists with a view to encouraging both to find new ways of communicating and shifting the public perception of environmental responsibility.

Chapter 10 (the legal chapter of this book) was a team effort, with contributions from the following individuals at Pinsent Masons, spanning diverse legal practice areas: Becca Aspinwall, Anne-Marie Friel, Rob Morson, Mark Marfe, Angelique Bret, Paul Williams, and Tadeusz Gielas.

Tom Glover, a self-proclaimed 'tech junkie,' has worked in the IT sector for over 25 years, starting with software application and technology layers such as machine learning, early-day AI, business rules, and algorithmic trading platforms before progressing to TCM and core banking applications. In 2009, he was lured to the dark side of data centres by David Ruberg with the role of leading Interxion's international business development. During his tenure, he and the team grew the business and share price by over 300%.

Over the past 12 years, Tom has overseen and steered data centre transactions worth over $10 billion and worked with occupiers, hyper-scalers, developers, landowners, governments, and institutional investors in this dynamic sector.

In his current role at JLL, Tom leads their EMEA data centre transactional business line across EMEA. In a recent poll (+170 replies), he observed that sustainability in the data centre arena was the number 1 challenge, with the resource (human capital) a close second. Tom believes that the data centre community is responsible for ensuring quality products are combined with best practices in achieving sustainability and bringing as many 'tech junkies' into the sector as possible. In a recent panel discussion, he said, 'The digital economy is not growing, it's exploding, and data centres are the bedrock foundation needed to support this!'

Tom Glover is currently DC Head of Transactions, EMEA, for JLL.

1

Introduction

As prefaced, this book is written primarily for lawyers, financial and insurance advisors, surveyors, engineers, architects, and other professionals who do not necessarily have direct experience with data centres but need to participate in the subject matter. The intention is to provide the reader with the broad landscape of technical and commercial issues and help identify a high-level overview of critical considerations. We trust that the book should provide the key concepts and terminology to complement their core skills when drafting documents or making business decisions and help them ask more informed questions when working on data centre projects. Data centres are a complex ecosystem with many stakeholders from different disciplines interacting at different phases of the facility's lifetime. The objective of this text is to provide the reader with a full spectrum of the entire life cycle of a data centre from inception to operational use.

Vincent Fogarty and Sophia Flucker have written the book, with Max Schulze and Andrew McMillan sharing their expertise in specific chapters.

Vincent Fogarty is a dual-qualified engineer and quantity surveyor who has acted as the appointed quantum and technical expert in mechanical and electrical matters in litigation and alternative dispute resolution processes in the United Kingdom and overseas. He has commercially managed various data centre projects, solved delay and cost disputes and has been an equity partner with data centre developments.

Sophia Flucker has specialised in data centres since 2004, working as a mechanical engineering consultant. Sophia's experience included secondments with operations teams, which broadened her practical knowledge helping data centre clients optimise their reliability and energy efficiency performance. Passionate about sustainability, she has collaborated with colleagues on several technical papers on data centre energy efficiency and environmental impact. She enjoys collaborating with different stakeholders to deliver award-winning work.

Max Schulze has a background as a software engineer and cloud expert, bringing his experience from start-ups and corporates to transforming the digital economy, making it future-proof and sustainable. He believes that sustainability is an opportunity and contributes to unlocking it with his work.

Andre McMillian is an international telecoms media and technology lawyer; he has a wealth of experience advising corporates, investment banks, and private equity houses on in-sector mergers and acquisitions and high-value commercial contracts.

Chapter 2 starts by examining data centres' central role in the digital economy and the growth trends due to the world's increasing reliance on data. In Chapter 3, whilst examining the multiple competing criteria layers of site selection, it is evident that the perfect data centre site may not exist. The optimal site criteria involve trade-offs that require an engineering solution to resolve the site-specific opportunities and risks.

In Chapter 4, Max demystifies computing and global connectivity, explaining how IT services work. This includes a macro architecture overview of all the connecting parts from data centres to fibre cable, terrestrial internet, Wi-Fi, and all the interconnecting infrastructure parts that ultimately connect the end users.

Sophia describes the key components of a data centre's physical infrastructure and various engineering solutions for cooling and powering the data centre in Chapter 5.

The reader is provided with the essential considerations and challenges of designing and building these complex projects in Chapter 6, including the project process, client requirements, design, installation, commissioning and handover, programme, budget management, what can go wrong, and lessons learned.

After the construction phase, the facility starts its operational life. Sophia explores considerations and challenges for cost-effective and efficient operation in Chapter 7, including faults and failures and the role of maintenance, management, and training to avoid and mitigate costly errors.

In Chapter 8, Vincent explains why data centres attract investors and the valuation considerations applied. There are many models by which finance may enter the data centre domain, and this book explores those fundamentals, including why data centres have a utility-like asset class that has the potential to provide a predictable income stream and therefore are investment grade for sovereign and private equity.

Data centres have a high and growing environmental impact due to their energy consumption and use of resources. Sophia addresses this in Chapter 9, which describes how engineering options and opportunities can mitigate and somehow avoid the worst impacts of these energy-intense operations.

In Chapter 10, Andrew explores the jurisdiction, regulation, and legal matters that wrap around the life cycle of data centre transactions and also looks at case law on when things go wrong and require a dispute resolution process. It is paramount to properly present and consider data centres in the relevant terms of the business.

Finally, in Chapter 11, because the data centre industry is full of innovation and is constantly evolving, the authors examine what may be around the corner and future trends in the industry, including regulation.

We hope for many that this book assists in identifying the unknowns and may provide a valuable source of continuous reference for the navigation of the data centre world.

2

What Drives the Need and the Various Types of Data Centres

From the outside, data centres may be seen as clean, windowless warehouses with thousands of circuit boards racked and arranged in rows, stretching down halls so long that operational staff on scooters may ride through the corridors of these rows of racks. But what drives the scale and growth of these intelligent data centre buildings? This chapter looks at the trends, current market, and industry structure and provides insights into the potential growth drivers, restraints, risks, and prospects. Those recent and legacy drivers of the data centre growth, the impacts, and the various generic types of data centres are all considered.

We, the populace, all interact with the internet, simply uploading our latest photos to Instagram, which may end up stored inside a vast data centre. The key driver's data centres that enable growth are becoming more recognisable. Data centres are where the cloud lives,[1] and our data, photographs, and music are stored. They are a critical component of the worldwide economy, whether you are a human being, a city, or a country. Likewise, the demand for data to enhance business performance is driving the growth of the data centre industry.

Data centres are also a fundamental part of the business enterprise, designed to support applications and provide services such as data storage, access management, backup, and recovery. Data centres also provide productivity applications, such as online meeting portals, e-commerce transactions, and provisioning for online gaming communities. Recently big data, machine learning, and artificial intelligence (AI) have prompted the growth of data centres.

---

1 Chai, W. (2021). *What is cloud computing? Everything you need to know.* [online] Search Cloud Computing. Available at: https://www.techtarget.com/searchcloudcomputing/definition/cloud-computing.

Data Centre Essentials: Design, Construction, and Operation of Data Centres for the Non-expert, First Edition. Vincent Fogarty and Sophia Flucker.

Cloud computing is a primary driver of data centre growth.[2] The cloud relies upon the pooling of data stored and then processed within the capabilities provided by the likes of Apple, Microsoft, Amazon, and Google. Users connect via internet devices, and through the network's tentacles, data centres allow users access to the data they need. The data is in all formats, from audio files, photographs, and computing software. Data centres are the internet's core, and the cloud is only made possible by high-speed, resilient, and reliable networks. These cloud networks may be public, private, or commercial.

Following the rise of the Internet of Things (IoT) and Industry 4.0,[3] manufacturers depend on big data analytics to enhance their operations' output efficiency and cost-effectiveness. The IoT usually refers to the instrumented world where IP[4] addresses are embedded in objects in the environment.[5] These 'Things' are devices operated in their home or carried by people. Modern-built assets tend to have intelligent doors, lighting, and controls that all interface with IP addresses. All types of Bluetooth, RFID,[6] GPS,[7] vehicles, and many more 'Things' are connected by the network's tentacles. The potential of a digital twin[8] that augments the creation of virtual reality offers the possibility to simulate all types of asset design and function scenarios, create extensive data, and compute demand.

Many IoT missions may require several locations for IoT data analysis and storage, including endpoint devices with integrated computing and storage; nearby devices that perform local computation; intelligent gateway devices; and on-premises data centres, managed to host sites, colocation facilities, and network providers' point-of-presence locations. The diversity of edge-computing locations reflects the diversity of markets for IoT.

---

2 Markets, R. (2020). *Comprehensive data centre market outlook and forecast 2020–2025.* [online] Globe Newswire News Room. Available at: https://www.globenewswire.com/news-release/2020/02/13/1984742/0/en/Comprehensive-Data-Centre-Market-Outlook-and-Forecast-2020-2025.html [Accessed 4 Sep. 2022].

3 Misra, S., Roy, C., Mukherjee, A. and Taylor (2021). *Introduction to Industrial Internet of Things and Industry 4.0.* Boca Raton, London, New York: Crc Press Is An Imprint Of The Taylor & Francis Group, An Informa Business.

4 An Internet Protocol address is a numerical label such as 192.5.2.1 that is connected to a computer network that uses the Internet Protocol for communication.

5 Greengard, S. (2015). *The Internet of Things.* Cambridge, Massachusetts: Mit Press.

6 AB&R (2019). *What is RFID and how does RFID work? – AB&R®.* [online] AB&R. Available at: https://www.abr.com/what-is-rfid-how-does-rfid-work/.

7 www.merriam-webster.com (n.d.). *Definition of GPS.* [online] Available at: https://www.merriam-webster.com/dictionary/GPS.

8 Nath, S.V., van Schalkwyk, P. and Isaacs, D. (2021). *Building Industrial Digital Twins Design, Develop, and Deploy Digital Twin Solutions for Real-World Industries Using Azure Digital Twins.* Birmingham: Packt Publishing, Limited.

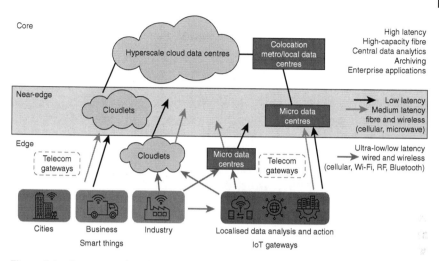

Figure 2.1 Data centre interfaces with the Internet of Things (IoT).

Several IoT deployments may end up storing, integrating, and moving data across a combination of public cloud and other commercial facilities, including colocation sites, with both distributed micromodular edge data centres and enormous centralised core data centres, including those of public cloud providers playing a role. Even within similar IoT applications, network architectures and data centre types may have various interfaces and data exchange paths, as shown in Figure 2.1.

The internet has primarily fuelled this sustained growth of data creation. The smartphone has been a big part of this growth. However, more IoT devices have further generated data through internet connections. The processing of mega quantities of data prompts the need for the internet via cloud computing because standalone technology does not have the capacity. The pivotal engine of this physical cloud computing infrastructure are data centres.

In this age of data, reports[9] indicate that there were 36 billion IoT devices installed worldwide by 2021 and a forecast of 76 billion by 2025. The generation of large masses of data affects the transactions to be captured, transmitted, stored, evaluated, and retrieved. Data centres house these treasuries of this internet age.

9 Statista (2012). *IoT: Number of connected devices worldwide 2012-2025 | Statista.* [online] Statista. Available at: https://www.statista.com/statistics/471264/iot-number-of-connected-devices-worldwide/.

The stock market confirms that some of the ten biggest global companies by market capitalisation are Alphabet,[10] Apple, Amazon, Microsoft, and Meta.[11] It may well be obvious how much data those big five produce and how it drives the data centre's needs. It may be less obvious that your local shop and sporting bookmaker also has data centre needs generally catered for by a colocator data centre provider. However, some businesses have privacy concerns about client data, such as banks, insurance companies, health providers, and others, who continue to have enterprise data centres.

Data Demand versus Compute Efficiency

There is a competing axis around which data centre size continually evolves. The first part is chipset efficiency, the second is software efficiency, and the third is rack density.

As we, the consumers of the data exchanges, continue to produce and depend upon more transactions and records, the pile of data increases organically. According to projections from Statista,[12] the total amount of data created, captured, duplicated, and consumed globally is forecast to increase rapidly. Until 2025, global data creation is forecast to rise to more than 180 zettabytes.[13] You can assume the relationship between data demand and data centre size is directly related; however, three factors influence that connection.

The first influence is chipset efficiency; the latest generation of server processors delivers more workload than those engaged previously. Every new server technology generation has delivered a leap in efficiency across the board for the past 15 or so years. In this context, it is worth recognising Moore's law as a term used to refer to the opinion given by Gordon Moore[14] when, in 1965, he said that the number of transistors in a dense integrated circuit (IC) doubles about every

---

10 Alphabet Inc. is an American multinational technology conglomerate holding company headquartered in Mountain View, California. It was created through a restructuring of Google on 2 October 2015, and became the parent company of Google and several former Google subsidiaries.

11 Meta as the overarching company that runs Facebook, WhatsApp, Instagram, and Oculus, among others.

12 Statista (2021). *Data created worldwide 2010–2025 | Statista*. [online] Statista. Available at: https://www.statista.com/statistics/871513/worldwide-data-created/.

13 Murphy, J., Roser, M. and Ortiz-Ospina, E. (2018). *Internet*. [online] Our World in Data. Available at: https://ourworldindata.org/internet.

14 Gianfagna, M. (2021). *What is Moore's law? | Is Moore's law dead? | Synopsys*. [online] www.synopsys.com. Available at: https://www.synopsys.com/glossary/what-is-moores-law.html.

two years. In 2021, Intel claimed[15] that the semiconductor industry would meet or beat Moore's Law.[16]

Second, some computer scientists point out that the efficiency or performance of the software decreases when the hardware becomes more powerful.[17] Many reasons are impacting this condition. The really significant reason is that the cost of creating software is dramatically increasing while, at the same time, computer hardware is becoming less expensive.[18] In 1995, Computer Scientist Niklaus Wirth stated, 'Software is getting slower more rapidly than hardware becomes faster'. This statement was later recognised as Wirth's Law.[19] It is because software comes to be more intricate as the hardware progresses; the actual execution of computers might not be improved as people anticipated. The term 'Software bloat' was created to describe the occurrence. Subsequently, computer scientists continued to make similar statements about Wirth's Law. British computer scientist Michael David May stated, 'Software efficiency halves every 18 months, compensating Moore's Law'.[20] This declaration became identified as May's Law. Whilst Wirth's Law, May's Law, and other laws contend that inefficiency counteracts the effect of Moore's Law, it is accepted that hardware efficiency trumps software inefficiency in productivity gains.[21] On the one hand, the software is slow and inefficient; on the other hand, the hardware industry follows Moore's Law, providing overabundant hardware resources.

The third is rack density within the data centre space. Racks are like a framing system that organises the high-density blade[22] servers,[23] network, and storage

---

15 VentureBeat (2021). *Intel promises industry will meet or exceed Moore's law for a decade.* [online] Available at: https://venturebeat.com/business/intel-promises-industry-will-meet-or-exceed-moores-law-for-a-decade/ [Accessed 2 Sep. 2022].

16 VentureBeat (2021). *Intel promises industry will meet or exceed Moore's law for a decade.* [online] Available at: https://venturebeat.com/business/intel-promises-industry-will-meet-or-exceed-moores-law-for-a-decade/.

17 See Chapter 6 of this book.

18 Luo, L. (n.d.). *Software efficiency or performance optimization at the software and hardware architectural level.* [online] Available at: https://core.ac.uk/download/pdf/38098467.pdf.

19 techslang (2021). *What is Wirth's law? — definition by Techslang.* [online] Techslang — Tech Explained in Simple Terms. Available at: https://www.techslang.com/definition/what-is-wirths-law/ [Accessed 3 Sep. 2022].

20 Wikipedia (2021). *David May (computer scientist).* [online] Available at: https://en.wikipedia.org/wiki/David_May_(computer_scientist) [Accessed 3 Sep. 2022].

21 published, C.M. (2012). *10 laws of tech: The rules that define our world.* [online] TechRadar. Available at: https://www.techradar.com/news/world-of-tech/10-laws-of-tech-the-rules-that-define-our-world-1067906 [Accessed 3 Sep. 2022].

22 SearchDataCentre (n.d.). *What is a blade server?* [online] Available at: https://www.techtarget.com/searchdataCentre/definition/blade-server.

23 Blade, cabinets, or server racks.

equipment. Each blade has an energy-consumed measure that may be stated in kilowatts (kW). The summation of power consumed in a single rack may range from 2 to 20 kW and sometimes beyond. This number of kW per rack is generally known as the rack density; the more kW per rack, the greater the density. The rack power density calculation is one of the most fundamental regarding server room and data centre designs. When the client or developer has decided on the data centre's capacity, the design density of the racks may offer most of the answer to the size of the floor area of the data centre. While densities under 10 kW per rack remain the norm, deployments at 15 kW are typical in hyperscale facilities, and some are even nearing 25 kW. An increased rack density for a total design load effectively reduces the data centre's footprint.

The development tendency has been to provide increased rack density and an ever-increasing chipset efficiency, thus getting more data transactions per footprint unit area of the data centre. In addition to floor area requirement, the power per rack multiplied by the total number of racks in the room provides the basis for capacity planning, sizing, critical power protection, and cooling systems. The industry trend is to squeeze more 'compute[24]' out of less footprint and power consumed.

It follows, therefore, that the chipset efficiency and power consumed to provide the data transaction is an ever-evolving equation and is a counterbalance to the ever-increasing demand for more data transactions. As the network latencies improve with the enabling of fully immersive virtual worlds that are accessible to everyone, the compute infrastructure layer continues to be pivotal in that journey. This increase in chipset efficiency may lead to more extensive retro refitting of data centres where new racks with new computer chipsets are replacing their older incumbents and reducing the need to build new data centres. However, the cooling requirements also increase as you increase the power load to higher densities. In retrofit projects where rack density increases the power usage and, therefore, the cooling need, it may prompt a complete redesign of the mechanical and electrical services with plant and systems replacement.

Artificial Intelligence (AI) has categories that include knowledge management, virtual assistants, semi-autonomous vehicles, virtual workplace, and machine learning. As networks become more complex, distributed, and augmented and virtual reality demands of the metaverse become more evident, the need for real-time computing and decision-making becomes more critical. The application and potential applications are so vast that I only touch upon those more obvious applications in the following paragraphs.

---

24 Network routing, computation, and storage.

The possibility for AI to drive revenue and profit growth is immense. Marketing, sales, and customer service were identified as functions where AI can realise its full potential. Using algorithms to improve the basics of account and lead prioritisation and requirement, suggesting the content or sales activity that will lead to success, and reallocating sales resources to the places they can have the most impact.

Some A1 was developed for vehicle autonomous on-road and off-road driving restrictions without driver interference. The autonomous driving system comprises four parts: localisation, perception, planning, and control. Localisation, which estimates the ego-vehicle position on the map, should be first performed to drive the A1 autonomously. A perception algorithm that detects and recognises the objects around the ego-vehicle is also essential to prevent collisions with obstacles and road departure. A planning algorithm generates the drivable motion of the A1 based on previous information from the localisation and perception system. Subsequently, a vehicle control algorithm calculates the desired steering, acceleration, and braking control commands based on the information from the planning algorithm. AI recreates human perceptual and dynamic cycles utilising deep learning algorithms and controls activities in driver control frameworks, like steering and brakes. The vehicle's software may counsel Google Maps for early notification of things like tourist spots, traffic signs and lights, and other obstacles. This vehicular real-time need is sensitive to ultra-low latencies,[25] and under the increasingly common hybrid model of enterprise, public and private clouds, colocation, edge AI, and machine learning will be critical to optimising the performance of these vehicles.

Health-related AI applications analyse the relationships between clinical techniques and patient outcomes. These AI programs are applied to diagnostics, treatment protocol development, drug development, personalised medicine, and patient monitoring and care. These applications have increased exponentially during the COVID-constrained period. Similar tranches of application are to be found in the teaching profession.

The application and potential of AI are vast and have been applied to manufacturing robots, self-driving cars, smart assistants, health care management, disease mapping, automated financial investing, virtual chatbot travel booking agents, and social media monitoring. Some might say that the only difference between humans and AI is that AI is a digital build on a binary system but humans are analogous with strands of emotional intelligence. The depth of this debate is nebulous, and somewhere in the realms of space odyssey, however, it would be hard to imagine any application of human endeavour that AI could not be applied.

---

25 Ashjaei, M., Lo Bello, L., Daneshtalab, M., Patti, G., Saponara, S. and Mubeen, S. (2021). Time-sensitive networking in automotive embedded systems: State of the art and research opportunities. *Journal of Systems Architecture*, [online] 117, p.102137. doi:/10.1016/j.sysarc. 2021.102137.

Generative AI models such as ChatGPT have the potential to replace many tasks currently performed by human workers.

Blockchain and cryptocurrencies are disruptive technical innovations in the recent computing paradigm. Many cryptocurrencies[26] depend on a blockchain. Blockchain is the technology that enables, among other things, the existence of cryptocurrency. A cryptocurrency is a medium of exchange, such as the Euro or US dollar, but is a digital universe and uses encryption to control the creation of monetary units and to verify the transfer of funds.

These blockchains and the contents within them are protected by resilient cryptography,[27] ensuring that transactions within the network may not be forged or destroyed. In this way, blockchain technology allows a digital currency to maintain a trusted transaction network without relying on a central authority. For this reason, those digital currencies are thought of as 'decentralised' and exist on the web in a way we like we keep cash or cards in a physical wallet; Bitcoins are also stored in a digital wallet. The digital wallet may be hardware-based or web-based. The wallet can also be located on a mobile device, on a computer desktop, or be kept secure by printing the private keys and addresses used for access on paper.

The difference between a cryptocurrency and legal tender currency is most notable when it comes to the tax treatment as property and not as currency[28] in most jurisdictions. Taxable gains or losses can be incurred similarly to property transactions.

Bitcoin is the best-known cryptocurrency for which blockchain technology was invented. While Bitcoin was born in 2008, virtual coins are 'minted' by miners who buy specialised servers to crunch time-intensive computations in a growing blockchain that proves the validity of the new crypto coins. Cryptocurrency is a computationally intense activity that increasingly needs more electricity and computing power. In addition to the high demand for computer hardware, the demand for data centres to accommodate this need is also rising exponentially.[29] As Bitcoin-mining equipment grows complex, the amount of heat generated by the hardware also increases and many solutions on how to handle the cooling demands of this intense hardware have emerged.[30]

---

26 Bitcoin, Ether (Ethereum), Binance Coin, XRP (Ripple), Tether and Dogecoin.

27 www.ibm.com (n.d.). *What is blockchain security? | IBM*. [online] Available at: https://www.ibm.com/uk-en/topics/blockchain-security.

28 Comply Advantage (2020). *Cryptocurrency regulations around the world*. [online] Comply Advantage. Available at: https://complyadvantage.com/insights/cryptocurrency-regulations-around-world/.

29 www.dataCentres.com (n.d.). *Cryptocurrency is changing the data centre market*. [online] Available at: https://www.dataCentres.com/news/cryptocurrency-is-changing-the-data-Centre-market [Accessed 4 Sep. 2022].

30 See Chapter 5 of this book for cooling solutions.

Given exceptionally high energy use and carbon emissions related to Bitcoin-mining operations raise anxieties about their effects on the global environment, local ecosystems, and increased consumer electricity costs. It is clear that crypto mining will significantly impact the data centre industry, although how far-reaching that is will remain to be seen.

Cryptocurrency mining is currently supporting the demand for data centre services, leading to a widening range of investment opportunities. Regulators are also interested in where the power is being sourced. The regulatory interest may drive increased adoption of renewables across the data centre industry as cryptocurrency-mining companies strive to lessen the environmental impact of their activities. And because crypto-mining processes run hot, it could also lead to increased investment in newer cooling technologies, such as liquid or liquid immersion cooling.[31]

Blockchain contracts are sometimes termed smart or Ricardian contracts.[32] These digital contracts are stored on a blockchain that is automatically executed when predetermined terms and conditions are delivered. These contracts are designed to automate an agreement's execution so that all participants can be immediately sure of the outcome without any transaction time lag. These executable codes run on top of the blockchain to enable, execute, and implement an agreement between untrustworthy parties without the participation of a trusted third party. They may also enable workflow automation, triggering the following action when conditions are met.

Smart contract applications include financial purposes like trading, investing, lending, and borrowing. They can be used for gaming, health care, and real-estate applications and can even configure entire corporate structures.

Colocation companies remain eager to cater to the blockchain community.[33] They continue to attract businesses needing colocation space for operating block-chain mining that typically involves compute-intensive software hosted on specialised hardware and consumes immense amounts of electricity. The hardware also manages to be quite loud and exhausts a lot of heat. These characteristics make blockchain-mining hardware a prime candidate for placement in colocation centres, where the equipment's owners may not have to worry about the speed of deployment, capacity, and cost. They may also benefit from the lower energy costs that colocation providers may sometimes offer.

---

31 Ibid.

32 Elinext (2018). *Smart vs. Ricardian contracts: What's the difference?* [online] Available at: https://www.elinext.com/industries/financial/trends/smart-vs-ricardian-contracts/.

33 Thin-nology (2021). *Does colocation have a place in the crypto world? – Thin-nology.* [online] Available at: https://thin-nology.com/does-colocation-have-a-place-in-the-crypto-world/ [Accessed 4 Sep. 2022].

Whilst new paths for growth may be prompted by blockchain and AI, legacy and new cloud demand are still being catered for by hyperscalers like Google, Amazon, and Microsoft, developing data centres for their own public clouds. It may be that the cloud is not for all organisations. At a particularly large scale, it may be more economical to keep data in-house rather than put it in the cloud. Some businesses in heavily regulated industries sometimes find it onerous to navigate compliance requirements involved in storing their data within a vendor data centre rather than in-house. For those companies like health care and finance, the security aspect remains an impediment for these companies. It may be that the cloud is reasonably secure, but there may sometimes be trust issues in turning over essential data to a third party.

Growing the data centre might remain the more attractive option for other types of businesses because of the workloads involved. The media,[34] gaming, and scientific industries might need to move terabytes of data around quickly, and there are trade-offs involved, even with the most robust cloud deployment that does not satisfy the need. It may be more pragmatic when working with vast amounts of data to keep that data locally rather than having a latency delay moving it back and forth from a cloud provider. In the data centre domain, one solution does not fit the actual need, and each business has a bespoke need and a counterpart solution.

The impact of the COVID-19-constrained years has been remarkable in many ways. A seismic shift has occurred in both usage patterns and scale of activity, with businesses moving operations to the cloud, a work-from-home workforce, remote learning through education providers, health care by remote consultation, etc., all mean that data centres are expanding faster.

Global capital expenditure on data centre infrastructure is set to grow by 10% over the next five years to a total of $350 billion[35] by 2026.

Edge computing[36] is forecast to comprise 8% of total data centre infrastructure spending by 2026. Edge computing is a dispersed computing framework that enables enterprise applications closer to data sources such as IoT devices or local edge servers. This proximity to data at its source can deliver substantial business benefits, including faster insights, improved response times, and better bandwidth availability.

---

34 aws.amazon.com (2019). *Migrating hundreds of TB of data to Amazon S3 with AWS DataSync | AWS Storage Blog*. [online] Available at: https://aws.amazon.com/blogs/storage/migrating-hundreds-of-tb-of-data-to-amazon-s3-with-aws-datasync/ [Accessed 31 Aug. 2022].

35 Dell'Oro Group (n.d.). *Global data centre capex to reach $350 billion by 2026, according to Dell'Oro Group*. [online] Available at: https://www.delloro.com/news/global-data-Centre-capex-to-reach-350-billion-by-2026/ [Accessed 31 Aug. 2022].

36 *One definition of edge computing is any type of computer program that delivers low latency nearer to the requests*. Also see: Al-Turjman, F. (2019). *Edge Computing: From Hype to Reality*. Cham, Switzerland: Springer.

Workload Placement

The decisions on workload placement are now more than just the physical selection of colocation, hosting, and cloud providers; there is a sustainable element to that selection process. However, when IT leaders are looking at workload placement based on business outcomes as a critical success factor, many considerations go into the decision process of where the workload is placed.

The decision process will be unique to each business type and a tranche of deliverables. There will be business considerations and top-down business problems that the organisation needs to solve, such as time to market, agility, and legal and regulatory. These business needs will have comparable technical factors like workload elasticity, data volume, integration, security, and performance are all complementary and, in some cases, competing priorities. Peer-to-peer system considerations include software as service (XaaS)[37] compatibility and cloud service provider (CSP)[38] alignment with business needs. There will also be generic considerations, including licensing, organisation practices, and disaster recovery.

The cloud may take a more significant percentage of the placement load, and this may be split into three primary parts: private cloud, public cloud, and hybrid. Private clouds may reside on-premises and may be managed internally or be located off-premises, managed by a third party and connected through virtual private networks (VPNs).[39] Public clouds are virtualised by computing, network, and storage resources offered and managed by a third party outside the customer's private network. Hybrid clouds provide a combination of both private and public cloud attributes and constraints.

Traditional placement is within an enterprise or on-premises data centre. These traditional data centres function as a legacy holding area, dedicated to specific services that cannot be supported elsewhere or supporting those most economically efficient systems located on-premises.

The cloud providers have introduced a range of Edge solutions, for example, Outposts from Amazon Web Services, Azure Stack Edge from Microsoft, and

---

37 SearchCloudComputing (n.d.). *What is XaaS (anything as a service)? – Definition from WhatIs.com.* [online] Available at: https://www.techtarget.com/searchcloudcomputing/definition/XaaS-anything-as-a-service.

38 SearchITChannel (n.d.). *What is a cloud service provider (cloud provider)?* [online] Available at: https://www.techtarget.com/searchitchannel/definition/cloud-service-provider-cloud-provider.

39 Wikipedia Contributors (2019). *Virtual private network.* [online] Wikipedia. Available at: https://en.wikipedia.org/wiki/Virtual_private_network.

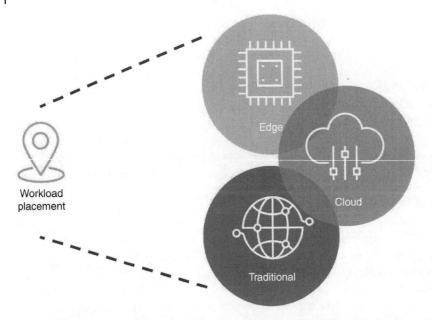

Figure 2.2 Workload placement; traditional on-premises or colocation, cloud, and edge.

Anthos from Google.[40] We have also seen that the vast amounts of data created by the IoT, specifically Industry 4.0, are driving the move to process data away from the cloud. Edge data centres are small data centres located close to a network's edge. They provide some of the same devices found in traditional data centres but are contained in a smaller footprint, closer to end users and devices. Ultimately, Edge Data Centres may eliminate the round-trip journey to the cloud and offer real-time responsiveness and local authority. Ultimately, infrastructure strategies may involve an integration of traditional on-premises, colocation, cloud, and edge delivery solutions, as outlined in Figure 2.2.

The merger and mix of these facilities will be unique to each business and will change over time as the needs change.

These workloads are placed in dynamic premises that facilitate and centralise an organisation's shared IT operations and equipment to store, process, and disseminate data and applications. Because they house an organisation's most important and proprietary assets, data centres are essential to the continuity of daily operations. Consequently, data centres' security, dependability, and information are among many organisations' top priorities.

---

40 Analyst firm Gartner, forecast that 'By 2022, more than 50% of enterprise-generated data will be created and processed outside the data centre or cloud, up from less than 10% in 2019'.

Historically data centres were controlled via physical infrastructure, but the public cloud has changed that model. Except where regulatory limitations require an on-premises data centre without internet connections, most of today's data centre infrastructure have evolved from on-premises physical servers to virtualised infrastructure that supports applications and workloads across multi-cloud environments.

The data centre industry has risen to meet the data generation use across a range of industries, which has increased demand for data servers and data centres. Demand for data centres differs by country, reflecting a variety of market forces and policies. For some jurisdictions, higher numbers of data centres may reflect industry-sector demand. The United Kingdom, for example, has the second-largest share of data centres and is one of the world's largest financial centres. The United States and China (first and fourth, respectively) have significant data demands across a variety of sectors, while the third largest, Germany, has a significant manufacturing and industrial capacity with enormous data demands.[41]

The ability of the internet to spread ideas across decentralised networks has prompted a political response in many parts of the world.[42] Government policies may direct the location of data centres in their jurisdictions with data localisation measures. These measures legally mandate that certain or all of the data of individuals in their jurisdiction be held within that jurisdiction. This may cover narrow classes of information like health data or a much more comprehensive range of information types. Russia, China, Turkey, Australia, Germany, France, and other countries have data localisation requirements with broad scope and enforcement variations.[43] The Russian and Chinese data localisation laws, in particular, are extensive and require extensive data to be held in domestically located servers.[44] Non-domestic firms have directly cited[45] several of these countries' data localisation measures as the justification for locating data within the country that

41 Daigle, B. (n.d.). *Data centres around the world: A quick look.* [online] Available at: https://www.usitc.gov/publications/332/executive_briefings/ebot_data_Centres_around_the_world.pdf.

42 O'Hara, K. and Hall, W. (2021). *Four Internets: Data, Geopolitics, and the Governance of Cyberspace.* New York: Oxford University Press USA – OSO.

43 In Country (2020). *Data residency laws by country: An overview.* [online] Available at: https://incountry.com/blog/data-residency-laws-by-country-overview/.

44 Internet Society (n.d.). *Internet way of networking use case: Data localization.* [online] Available at: https://www.internetsociety.org/resources/doc/2020/internet-impact-assessment-toolkit/use-case-data-localization/ [Accessed 3 Sep. 2022].

45 Itif.org. (2022). *How barriers to cross-border data flows are spreading globally, what they cost, and how to address them.* Information Technology & Innovation Foundation. [online] Available at: https://itif.org/publications/2021/07/19/how-barriers-cross-border-data-flows-are-spreading-globally-what-they-cost/.

otherwise may not have been located there. For example, following the introduction of Turkey's data localisation law in 2021, several US firms (Facebook, Twitter, Google) announced that they would be locating Turkish data within Turkey. While LinkedIn opted not to operate in Russia due to a desire not to set up a data centre within the Russian Federation.[46]

The Core Components of a Data Centre

Data centre architectures and requirements can differ significantly. For example, a data centre built for a CSP like Amazon satisfies facility, infrastructure, and security requirements that may significantly differ from a private data centre, such as one built for a government facility dedicated to securing classified data.

Regardless of classification, an effective data centre operation is achieved through a balanced investment in the facility and the equipment it houses. In addition, since data centres often house an organisation's business-critical data and applications, facilities and equipment must be secured against intruders and cyberattacks.

Types of Data Centres

Not all data centres are the same, and it is worth considering the various types of data centres, including enterprise, public and private clouds, colocation, and edge-type data centres. Data centres may be divided into four main categories: Enterprise, Hyperscale, Colocation, and Edge.[47]

Entreprise Data Centre

Enterprise data centres are built, owned, and operated by companies and optimised for their customers. Most often, they are located within the premises of their own corporate campus. Enterprise data centres are a traditional local server set-up. Enterprise data is operated as its own service platform to companies that have chosen not to outsource their operational IT needs to, for example, a colocation or public cloud base solution. Thus, the enterprise data centre's servers may be typically located in a dedicated room in the company's own buildings and thus form the basis for the company's internal network and IT solutions. Some of the

---

46 Daigle, B. (n.d.). *Data centres around the world: A quick look.* [online] Available at: https://www.usitc.gov/publications/332/executive_briefings/ebot_data_Centres_around_the_world.pdf.

47 STL Partners (n.d.). *Edge data centres: What and where?* [online] Available at: https://stlpartners.com/articles/edge-computing/edge-data-centres/ [Accessed 3 Sep. 2022].

larger enterprise businesses like financial services may have dedicated standalone data centre buildings that appear as discrete as those simply within dedicated rooms within multi-functional offices.

A private cloud is a computing service offered over the internet but only for a single customer instead of the general public. It may also be described as a corporate or internal cloud. Data privacy is supported through firewalls and internal hosting. Third-party providers may be denied access to operational and sensitive data.

Cloud computing refers to the accessibility of computer systems and resources on the internet as an alternative to locally on your computer, such as storage. Businesses may benefit from cloud computing in terms of its scalability, availability, instant provisioning, virtualised resources, and storage. Cloud computing is divided into three subsets: private, hybrid, and public cloud computing. A hybrid cloud[48] refers to a combined computing, storage, and services environment. It may combine both private and public cloud computing. Public cloud[49] computing describes computing services offered by a third-party provider and shared with multiple organisations using the public internet. These offer various kinds of software, and their security is different.

Like all cloud computing, private cloud is off-premises, meaning data is stored and accessed online on the internet so that a user can gain entrance to it whenever necessary. The private cloud allows restricted use to a single customer meaning other people cannot access the data, unlike the public cloud, where anyone can access the data. The private cloud offers certain security advantages over the other two types of cloud computing, though it still poses some of the same security risks as the public cloud, such as outdated virtual machine images, untested administration service providers, and data loss. Service providers in the private cloud manage data maintenance.

There are similarities between a private cloud and a data centre. They are both paths through which data may be stored and made available to users. Both platforms may be maintained by trained IT experts, with a private cloud by service providers and a data centre by developers.

The distinctions between a private cloud and a data centre are that a private cloud is off-premises while a data centre is on-premises. A private cloud is off-premises because it is accessed from the internet. A data centre is on-premises as the applications are located near the organisation. A private cloud is easily

---

48 azure.microsoft.com (n.d.). *What is hybrid cloud computing – definition | Microsoft Azure.* [online] Available at: https://azure.microsoft.com/en-gb/resources/cloud-computing-dictionary/what-is-hybrid-cloud-computing/.

49 azure.microsoft.com (n.d.). *What is a public cloud – definition | Microsoft Azure.* [online] Available at: https://azure.microsoft.com/en-gb/resources/cloud-computing-dictionary/what-is-a-public-cloud/.

scalable, requiring only a tiny amount of investment, while a data centre is not easily scalable and needs a considerable investment of servers.

Though a private cloud is reasonably secure, its use still has security risks. It may be, therefore, not suitable for critical projects. On the other hand, a data centre is more secure and, therefore, more suitable for critical projects. It is commonplace for a business to have a hybrid approach with enterprise data centre use and other private and public cloud applications.

Hyperscale or public cloud are large data centres often owned, operated, and used by internet-only companies such as Microsoft, Google, Facebook, and Amazon. Hyperscale data centres often act as service providers, forms of social media, search engines, streaming services, and e-commerce sites.

The notable difference in how a hyperscale cloud differs from a private cloud is that a hyperscale cloud is usually a multi-tenant platform where computing resources can be accessed on-demand. On the other hand, private cloud hosting offers a single-tenant platform that runs on dedicated infrastructure.

Due to the online presence of these companies, this type of enterprise requires a large data capacity. Therefore, well-known companies such as Google, Facebook, and Apple often build these hyperscale data centres to achieve the necessary data capacity needed to maintain and develop their online business model.

Colocation

Colocation is the practice of housing client servers and networking equipment in a third-party data centre. Instead of the in-house scenario where servers are located within a room or a section of an organisation's business infrastructure, there is the option to 'colocate' equipment by renting space in a colocation data centre. It is a shared facility under the terms of a lease under which the customers share[50] the cost of power, communication, cooling, and data centre floor space with other tenants.

Colocation may suit businesses that require complete control across their equipment. Companies may maintain their own equipment the same way they do when servers are installed in-house. As an alternative to building a new data centre, businesses may simply augment their current data centre by employing the space in a colocation facility. Housing data hardware in a colocation-type data centre may give companies access to higher bandwidth associated with a standard office server room at a much lower cost.

---

50 Keystone Law (n.d.). *Service level agreements*. [online] Available at: https://www. keystonelaw.com/keynotes/service-level-agreements#:~:text=A%20Service%20Level%20 Agreement%20(or.

Public Cloud

There is a significant yet discrete distinction in how a colocation compares with the public cloud. The distinction between colocation and the public cloud is how the data is stored and managed. It is a matter of having physical assets as opposed to virtual ones. Like colocation, cloud-based infrastructure services offer cost savings because of shared facilities. Many businesses prefer cloud-based services because they select to use their own data centres for more valuable tasks in expanding the business. Others choose a cloud provider because of the flexibility of rapidly scaling data capacity up or down based on fluctuating business needs.

But the convenience of the cloud may have its downsides. Cloud providers may offer the benefit of managing business data, but the conflict lies whenever data expands. More data means additional storage and costs. This route is more adaptable because businesses only rent the space for all assets, which is distinct from renting the asset itself, the cloud storage, which will be limited based on the subscription.

Urban or Edge

Many would say that Edge computing will drive future data centre's development. Edge computing permits data to be processed closer to the source, reducing latency. It is also worth taking into consideration the relationship between edge computing and the 5G network as an alternative to classical data centres, allowing faster development of AI or IoT applications.

5G is upgrading the current 4G network to facilitate data-heavy smartphone use. 5G may increase speeds by up to ten times that of 4G, while mobile edge computing reduces latency by bringing compute capabilities into the network nearer the end user. The purpose of 5G is to galvanise IoT by supporting the decentralisation of data processing by IoT, now termed edge processing. This edge processing prompts the need for edge data centres, smaller facilities near the populations they serve that deliver cloud computing resources and cached content to end users. They characteristically connect back to larger central data centres or multiple data centres.

We may also see the development of mobile data centres, which may be easily situated, for example, near significant sports events, concerts, etc. Edge data centre solutions may also eliminate connectivity problems in remote areas.

According to statista.com, the value of the global intelligent home market will rise to $222.90 billion by 2027.[51] As more people adopt smart solutions, not only

---

51 Statista (n.d.). *Smart Home – Worldwide | Statista Market Forecast.* [online] Available at: https://www.statista.com/outlook/dmo/smart-home/worldwide [update: Dec 2022].

within employment locations but also in their homes, the requirement for edge computing will keep growing, and so will the demand for edge computing devices. This will change data centres as the decentralisation of computing power will directly impact the type of data centre we will see in the future. In effect, data storage and processing may be placed in remote locations with medium and high latency and the networking and compute function nearer the customer side transaction with low latency.

This edge urban data centre is a smaller type of data centre that serves customers with a demand for data capacity with very low latency. This type is evolving in line with the needs of a growing customer base that demands edge computing. Due to the necessity for very low latency, this type of data centre is required to be near its customers. This shows this type of data centre, often in cities or suburbs. Urban data centres are increasingly crucial in deploying 5G and IoT due to an increased need for data centres to be located close to the end user.

As the urban data centres are more diminutive than hyperscale cloud data centres, there is typically an electrical power of 50–400 kW per data centre, corresponding to the electricity consumption of between 438 and 3504 MWh per year. Due to the urban nature of the data centres, these are typically connected to the local electricity distribution network without the need for new electrical substation infrastructure because of their lower power demand.

Urban data centres are often designed and constructed as containers or modular systems with small mechanical refrigeration systems.

The data centre industry is critical to the twenty first-century life, and its importance will increase concurrently with growth in and reliance on digital communication services. With more people accessing the internet, with more businesses moving to online product and service provision, with the growth of the IoT and connected devices, and with the desire to create, store, and analyse data in more significant volumes, faster, and in a more universally accessible manner than ever before, the already-exponential growth in data over recent years is just the beginning of a revolution whose impacts will be felt globally.

3

Site Selection

Choosing a location for a data centre requires being sensitive to the business and the surrounding community's needs for a long-term investment. To select a site to create a data centre, you ought to consider and evaluate many variables, from uncovering desirable physical characteristics and avoiding possible hazards to locating in an environment favourable to business growth and development.

Frequently, we hear about current and potential clients looking to develop large-scale data centre campuses. Given the immense expected growth in data centres to support a domestic and international surge in edge computing, artificial intelligence (AI), content creation, and an even more significant increase in digital learning and remote work, this is no surprise. Frequently, these developers have a unique piece of land in strategic geography, abundant natural resource traits, climate or temperature-related benefits, a favourable agreement with the site's municipality, and immediate access to significant power sources. It is essential to assess these properties from various perspectives, including understanding the requirements of cloud, colocation, and build-to-suit operators; designing, commissioning, and migrating to large-scale data centres for enterprises and institutions; and considering a sustainable design.

Site selection and development is quite a competitive space; developers are all looking for the advantage of making their campus more appealing to large-scale end users. Currently, most large data centre operators are intensely engrossed in energy usage, greenhouse effects, and sustainability driven by government directives and their own corporate objectives. Some of these projects are exploring both completely off power grid solutions or exploiting the most efficient and pioneering technologies to drive a sustainable operation.

Building a data centre is a mission-critical endeavour, from articulating the business need to defining the critical elements of an appropriate site. In order to execute a data centre's vision successfully, it is necessary to create a detailed

Data Centre Essentials: Design, Construction, and Operation of Data Centres for the Non-expert,
First Edition. Vincent Fogarty and Sophia Flucker.
© 2023 John Wiley & Sons Ltd. Published 2023 by John Wiley & Sons Ltd.

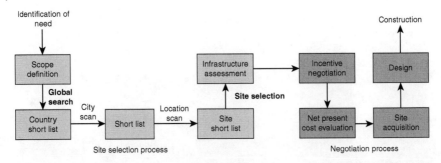

Figure 3.1 Site selection to a construction decision.

picture of the crucial and complex requirements, which interface with the geographical location and assess any potential 'red flags' that would automatically disqualify a property.

Site selection in the data centre industry is complex and encompasses multiple competing criteria layers. While accepting that the perfect data centre site may not exist, challenging analyses of all the relevant site selection criteria are necessary and will ultimately involve inevitable trade-offs in reaching a final decision to select and procure a site to start construction. Starting with a well-defined site selection process with the essential steps, as outlined in Figure 3.1, may enable the quantitative analysis of factors that translate unrefined data into helpful information and insight, facilitating informed decision-making.

Every potential site has advantages and trade-offs, but with modern construction innovations, you can mitigate many of them. An engineering solution can overcome many challenges presented by a site location. For example, a site located in a hurricane zone might feature a hardened exterior exceptionally resistant to natural or physical threats like excessive wind speeds. Early identification of crucial issues helps mitigate risks and enables building in a challenging location. Every mitigation solution will present a cost-benefit analysis that may impact capital or operation costs. Selecting a site from many potential options may present a grading system that will help in providing the right optics for the ultimate selection decision, as per Figure 3.2.

There are perhaps two limbs of approach to selecting a site driven from different starting positions. In the first part, you may have the developer-type entity with local knowledge of a potential area and then develops a proposal to the data centre market and takes what may be termed a bottom-up approach.[1] These developers

---

1 The bottom-up approach is when a developer finds the site and then attracts the data centre investor(s)/client to procure or joint venture in developing the proposed site to build a data centre.

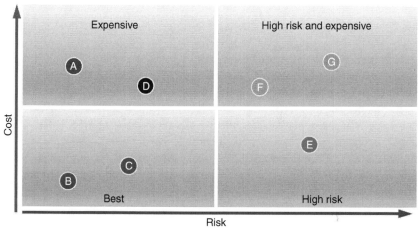

Figure 3.2 Cost = construction, power, fibre, cooling, labour, local taxes
Risk = seismic, weather, water, pollution, political, economic

Figure 3.2 Grading matrix for site selection.

often operate with private equity as the financial market continues to see a strong appetite for data centres from investors who are starting to view data centres in the same investment grade category as more traditional real-estate sectors. The equity investors may cover the site due diligence cost, design, and legal cost to the planning permission stage with the objective of a staged exit once an interested data centre operator engages. The legal title of the land is generally not transferred until the final operator is ready to procure. Option agreements and site assembly and development are the preferred paths of seed investors. An option to purchase[2] is an incredibly useful tool for a data centre developer seeking to assemble ownership of a site for a development scheme. Several different parties may own the land that comprises the intended development site. Using options to acquire these interests can provide the developer with the required combination of certainty and flexibility to exit if an end purchaser for the scheme does not materialise.

These investors may enter a joint venture special purpose vehicle-type company (SPV) with the site developer to attract end user clients before or after the grant of planning permission or licence stage. A waterfall model[3] is often used to umbrella multilabel parties in a shared investment. It is a term used in a Legal Operating

2 Reed, R. (2021). *Property Development*. New York: Routledge. A call option or sometimes a promotion agreement is exercised between the developer and landowner.

3 Realty Mogul. (n.d.). *Commercial real estate waterfall models for private placement offerings.* [online] Available at: https://www.realtymogul.com/knowledge-centre/article/waterfall-models-commercial-real-estate-equity [Accessed 14 Aug. 2022].

Agreement that describes how money is paid, when, and to whom in the development of equity investments. In short, once various predetermined milestones are achieved, the initial promoter may receive a more significant portion of the cash flows in proportion to their equity invested when the deal was initially acquired, also known as promoted interest. Each waterfall structure is different and requires considerable expertise.

The second part of sourcing a data centre site is the mega-scale entity with a global need and a regional focus, which peruses what may be termed a top-down approach.[4] Once these entities have identified their market need, they look at multiple locations within diverse jurisdictions. Many specialist estate agents are in the mix of these two binary parts, engaging with both entities to bind a deal.

To have a successful site that will mature into a development project and an operational data centre, one needs to consider and provide weighting to the criteria in selecting a site. All too often, expensive oversights and omissions in due diligence in Site selection become evident during the build and operational phase. The criteria outlined later in this chapter are generic and may have different weighting depending on the data centre's need.

Climate

Those weather conditions prevailing in a potential site area over a long period are vital considerations. The weather risk to a site is the convolution of three factors: hazard, exposure, and vulnerability. A site selection that avoids weather risks is always preferred. However, specific exposures and vulnerabilities may be acceptable risks that have engineering solutions.

It may be best to avoid a site that experiences natural hazards such as floods, poor air quality, earthquakes,[5] or volcanoes.[6] It is wise to consider severe historical weather events; for example, ancient stone makers in Japan left markers that warned against constructing anything below a particular elevation because of the risk of tsunamis, and those old warnings proved to be correct.[7] In considering any site, the flood risk must be assessed as water and electronics should not mix; appropriate geotechnical investigations are necessary.

---

4 The top-down approach is where the data centre need is identified and then the sourcing of the site follows that identified need.

5 Seismic vulnerability studies.

6 Volcanoes may be particularly problematic; whilst the initial volcanic eruption poses a local risk, the prevailing wind pattern may pose a prolonged risk.

7 Japan earthquake and tsunami of 2011, also called Great Sendai Earthquake or Great Tōhoku Earthquake, severe natural disaster that occurred in north-eastern Japan on 11 March 2011.

Access Roads and Airports

However, it is not only the avoidance of the data centre site becoming flooded; the access roads must be accessible so that staff and specialists may have the ingress/egress to the build phase and, ultimately, the operational data centre. Site access is essential; therefore, a site near a major highway is attractive for bringing in people and building materials with multiple roads to access a site; this lessens any concerns over possible human-made or natural disasters unexpectedly blocking access.

Climate may significantly affect the cooling efficiency of a data centre, which in turn affects operating costs. In general, dry, cold climates support the most efficient, cost-effective cooling solutions, although there is a more negligible difference than often perceived between temperature and cold climates.[8]

The local climate is a significant factor in data centre design because the climatic conditions determine what cooling technologies ought to be deployed. The selection of cooling systems impacts uptime and dramatically impacts a data centre's power costs. The different configurations and the cost of managing a data centre in a warm, humid climate will vary significantly from working in a cool, dry environment. Nevertheless, data centres are located in extremely cold and hot regions, with innovative methods used in both extremes to maintain preferred temperatures within the centre.

Water is often a critical resource for data centres because evaporation[9] is often the most cost-efficient cooling method. This method uses a lot of water. For example, on-site water consumption is estimated at 1.81 (0.46 gallons) per kWh of total data centre site energy use.[10]

Air Quality

Outdoor air used for ventilation, pressurisation, and cooling remains the primary source of airborne contaminants to data centres. However, data centres are dynamic environments where maintenance operations, infrastructure

---

8 See Chapter 5 of this book for cooling solutions.

9 See Chapter 5 of this book; adiabatic cooling is used in evaporative coolers. An evaporative cooler is basically a large fan that draws warm air through water-moistened pads. As the water in the pads evaporates, the air is chilled and pushed out to the room. The temperature can be controlled by adjusting the airflow of the cooler.

10 Shehabi, A., Smith, S., Sartor, D., Brown, R., Herrlin, M., Koomey, J., Masanet, E., Horner, N., Azevedo, I. and Lintner, W. (2016). *United States Data Centre Energy Usage Report.* [online] Available at: https://eta-publications.lbl.gov/sites/default/files/lbnl-1005775_v2.pdf.

upgrades, and equipment changes occur regularly, possibly introducing airborne pollutants.[11] Data centres also house other contaminants, such as chlorine, that can be emitted from PVC insulation on wires and cables if temperatures get too high.

Therefore, the air quality may affect the efficiency and operation costs of a data centre, equipment function, and employees' health. In high-pollution areas with fine particulates, such as those generated by diesel fumes, dust storms, or heavy pollen, the data centre's air supply may require more expensive carbon-type filters and more frequent filter changes adding to the initial and operational costs. It may be best to avoid locating sites near oceans as high sulphates and natural corrosive salts can damage circuit boards. Some data centres in urban locations have reported failures of servers and hard disk drives caused by sulphur corrosion. The International Organization for Standardization (ISO)[12] has developed guidelines that summarise acceptable contamination levels.

It is best to avoid sites close to heavily wooded areas or significant sources of airborne pollution; chemical pollutants such as automotive pollution and smoke from local fires can harm equipment.

Likelihood of Natural Disasters

A significant factor in locating a data centre is the stability of the actual site. Weather, seismic activity, and the probability of weather events such as hurricanes, fires, or flooding should all be considered.

The centre's location may be a more dominant factor, and the facility is strengthened to withstand anticipated threats. One example is Equinix's NAP of America's data centre in Miami, one of the largest single-building data centres, 750,000 square feet and 6 stories, which is built 32 ft above sea level and designed to endure category 5 hurricane winds. The Bahnhof Data Centre under White Mountain in Stockholm is at the other extreme of protection, located within the ultra-secure former nuclear bunker Pionen in Stockholm. It is buried 100 ft below ground within the White Mountains and secured behind 15.7 in.-thick metal doors.

---

11 And, R. (2014). *Particulate and Gaseous Contamination Books in Datacom Environments, Second Edition*. Ashrae Publications/American Society Of Heating, Refrigerating And Air-Conditioning Engineers.

12 www.iso.org (2016). *ISO 14644-1:2015*. [online] ISO. Available at: https://www.iso.org/standard/53394.html.

Ground Conditions

Almost like all forms of construction, the data centre's prospective site requires site-specific geotechnical investigation and the subsequent interpretation of data to determine the best structural solutions.

Ground conditions encompass a broad domain of risks but may include underground quarries, wells, and withdrawal-swelling of clays. The risk of the withdrawal-swelling of clay soils under the effect of drought withdrawal can be observed on clay soils, resulting in differential settlements that can sometimes cause severe damage to buildings.

However, the data centre site may require specific consideration[13] for the quality of the ground for earthing the electrical system.[14] Earth resistance generally depends on soil resistivity. It is difficult to achieve good grounding in sandy grounds and wood areas of high resistivity[15] as it requires far more resources than in wetlands with low soil resistivity.

Communications Infrastructure

The global economy progressively depends on telecom network systems. These networks join up the internet[16] of end users with providers via complex connections and terrestrial signals that transport data back and forth to and from data centres. Such connectivity is, in a sense, the *lifeblood*[17] of data centres.

Various transmission media types, such as fibre, copper, satellite, and microwave-like 5G,[18] connect all of the binary parts, which are used to give global connectivity. Physical media such as fibre or copper is preferred over satellite and

13 American, Telecommunications Industry Association, Electronic Industries Alliance, For, A. and Firm, D. (2002). *Commercial Building Grounding (Earthing) and Bonding Requirements for Telecommunications*. Washington, DC: Telecommunications Industry Association.

14 British Standards Institution (2020). *Telecommunications Bonding Networks for Buildings and Other Structures*. London: British Standards Institution, -03-05.

15 GreyMatters (n.d.). *Data centre earthing design case study of a large capacity data centre*. [online] Available at: https://greymattersglobal.com/case-studies/data-centre-earthing-design/ [Accessed 14 Aug. 2022].

16 Techopedia.com (2019). *What is the web? – Definition from Techopedia*. [online] Available at: https://www.techopedia.com/definition/5613/web.

17 Datacentremagazine.com (2021). *Data centres – the lifeblood of hybrid and remote working*. [online] Available at: https://datacentremagazine.com/data-centres/data-centres-lifeblood-hybrid-and-remote-working [Accessed 14 Aug. 2022].

18 Ericsson (2019). *5G – the 5G switch made easy*. [online] Ericsson.com. Available at: https://www.ericsson.com/en/5g.

microwave because both are susceptible to interference and atmospheric conditions. Most data centre connectivity is via fibre-optic cable.

Since data centres produce a significant amount of network traffic, they must have adequate fibre capacity for current and future needs, with at least the ability to add more fibre as demand grows. Fibre access is a significant asset when long-haul[19] and local fibre carriers are close, along with solid bandwidth. Potential site proximity to a subsea fibre landing station may well indicate long–haul potential to various continents. A due diligence fibre-optic report by an expert firm is always a worthwhile investment in the early stages of site selection.

Latency

Network latency, the nemesis of digital business, deterministically affects business performance and client satisfaction. Business transactions rely on applications running all aspects of business operations. Customers may need quick access to applications, and the applications increasingly need to communicate with each other. The further data must travel from where it is stored to where it is used, the higher the latency, so data centres may need to be located in proximity to business hubs to reduce latency issues depending on the type of application. The further a data centre is from a peering point, the greater the risk of connectivity issues. Therefore, data centres must be located in well-connected areas with access to many carriers and multiple redundant fibre connections to major bandwidth providers. The preferred way to provide consistent and reliable bandwidth at the volumes required by an enterprise-grade data centre is to build links to various network providers. So proximity to an urban hub is advantageous.

The most dominant connector of the internet is the fibre-optic cable. The cable used inside a building differs from that used outside a building. Inside a building, you typically use multi-mode fibre[20]; outside, you use single-mode fibre.[21] In considering site selection, the focus will be on single-mode fibre.

Fibre-optic connectivity fundamentally works by sending and receiving light pulses through a single-mode cable. The laws of physics govern the speed at which that pulse of light travels. Einstein determined the speed of light within a vacuum as 'C', or approximately 186,282 miles per second. The pulses of light in a trading scenario are through a piece of glass and are not travelling through a vacuum,

---

19 Mobasheri, A. (2001). *Computer Science and Communications Dictionary*. Boston, MA: Springer US.

20 Senior, J. (2013). *Optical Fiber Communications: Principles and Practice*. Ndorling Kindersley (India): Prentice-Hall.

21 Ibid.

effectively being slowed down by the manufactured specification of the glass fibre. The refractive index of light in a vacuum is one. The refractive index is the ratio between 'C' and the speed at which light travels in a material such as fibre.[22]

Latency is generally understood as the time between a command issuing and a server responding[23] in return trip time.[24] This may be, for example, an employee trying to interact with a document or a customer attempting to open a webpage. Latency has time impacts on every interaction you have with your server. The higher the latency is, the longer it takes, the greater the productivity loss, and the more frustration users experience. The state of the network compounds the round-trip time. Data rarely journey in a straight line between sender and recipient. Instead, it strolls through networks, routers, and switches, each of which can add latency. As a law of physics, the closer the data centre is to its customers, the lower the latency.

The refractive index for single-mode fibre can vary slightly based on several factors. However, a reasonable estimate is around 1.467, meaning that light travels through fibre at $186,282/1.467 = 124,188$ miles per second. As a result, you would save approximately eight microseconds if you could shorten your physical fibre route by one mile between exchange venues. Therefore, having the shortest path will always help minimise latency.

Because of the distance consideration, latency may be pivotal in Site selection and with the customer interface applications of the data centre.

Direct fibre paths from one point to another are preferable, with the least amount of network switching and splicing.[25] Most providers, whether dark, fractional, or managed, utilise diverse routes to get between geographical locations. As a result, you see different latencies. If you genuinely want the fastest connections between multiple venues, you will probably need to go with numerous providers and optimise the current network configuration and optimisation.

There is no logical solution to needing a speedy, low-latency transaction served to the customer from a grossly remote location. For example, financial firms in London's financial district are involved in high-frequency trading. The rapid-fire nature of their transactions means even milliseconds count. Having a data centre far away could put them at a disadvantage, even though digital signals travel at the

22 Senior, J.M. and Yousif Jamro, M. (2009). *Optical Fiber Communications: Principles and Practice*. Harlow, England/Toronto: Financial Times/Prentice Hall.

23 Reis, J. (2022). *Fundamentals of Data Engineering: Plan and Build Robust Data Systems*. Sebastopol: O'reilly Media.

24 Expressed in Return Time Trip (RTT) in milliseconds (μs); If the object to be illuminated is 100 metres from the flashlight and the latency from flashlight to object is 0.33 μs, the time required for the light to travel from the observer's flashlight to the object and back to the observer (round-trip delay) is 0.66 μs.

25 Yablon, A.D. (2023). *Optical Fiber Fusion Splicing. Springer Series in Optical Sciences*. Springer, p. 103.

speed of light. What internet gamers need is much like what high-frequency traders need. Their high-stakes buy-sell games, that is, high-performance and low-latency solutions, ensure the delivery of game functions to consoles and mobile devices with minimal lag time and the best gaming experience for users.

In the alternative there is no apparent benefit, in having a slow high latency transaction, near the customer. It is cheaper to move photons than electrons because electrical power networks require more physical cables and transformers to carry long distances. Hence, data centres are being built close to the power network substation point of connections to the available generation where low latency is not critical.[26] The application and usage will drive the combination of power and latency requirements, so not all scenarios are equal. High-latency applications may be suitable for remote data centre locations that involve data storage or cryptocurrency mining.

Latency will be affected by the data centre's proximity to your point of use and the nearest internet exchange point, with closer to both being the optimal position.

Low latency will either be vital or not necessary, depending on the types of applications. Since latency is a function of the speed of light, the distance between the computation and the transaction will effectively constrain the latency times. For example, in London's Canary Wharf, there is a relatively high density of data centres due mainly to the needs of the financial markets; on monetary exchange and share trading floors, milliseconds are vital.[27] However, if you are a Facebook user in Canary Wharf, you may well be looking back on a five-year-old picture stored in Meta's Lulea[28] or Odense data centres.[29] Low latency is not essential for many document storage types and streaming services like Netflix, where the back-end serving systems can be remote from the customer. The edge data centres with integrated 5G[30] capability will serve ultra-low latency needs such as AI and machine learning algorithms. These edge data centres will be relatively small in capacity.[31] They may act as the continuum of physical infrastructure that comprises the internet to and from centralised data centres to devices.

---

26 sdialliance.org (n.d.). *Home.* [online] Available at: https://sdialliance.org/ [Accessed 15 Aug. 2022].

27 LSEG (n.d.). *Exchange hosting.* [online] Available at: https://www.lseg.com/exchange-hosting [Accessed 15 Aug. 2022].

28 www.facebook.com (n.d.). *Luleå data centre.* [online] Available at: https://www.facebook.com/LuleaDataCentre/ [Accessed 15 Aug. 2022].

29 Meta (n.d.). *Odense data centre.* [online] Available at: https://about.fb.com/media-gallery/data-centres-2/odense-data-centre/ [Accessed 15 Aug. 2022].

30 Perry, L. (2020). *IoT and Edge Computing for Architects: Implementing Edge and IoT Systems from Sensors to Clouds with Communication Systems, Analytics and Security.* England: Birmingham.

31 200 kW–1 MW according to the 2022 State of the Edge The Linux Foundation.

Proximity to Subsea Cable Landing Sites

It may well be those undersea cables[32] that drive onshore site decisions. Some of the locations where those cables crawl ashore from the deeps of the sea are landing their share of data centre developments because these new lines increase the connectivity.

Increasingly, submarine cable projects are forgoing the cable landing station (CLS)[33] and connecting directly to data centres. Low latency with easy customer access and cost-saving are among the benefits of doing so. However, bringing the subsea cable further inland also has its share of risks. One risk is that the high-voltage direct current that drives amplifiers and repeaters makes the transit above or below ground challenging. It makes sense to land subsea cable with a pre-established data centre to serve the CLS and the point of presence (PoP)[34] role.

If a subsea cable is terminating at a data centre, you are at an internal cross-connect distance from the other internet traffic. It is simple to cross-connect[35] with the other customer on a different route. However, when you intend to traffic data via a stand-alone landing station, it is not as simple as that because customers do not have access to the landing station. So you need to pay for a backhaul network to connect a lot more problematic to do.

Density of Fibre Telecommunication Networks Near the Data Centre

Latency is the most important reason why consumers choose to leave mobile sites,[36] with 10% attributing slow downloads as a justification for not purchasing.

High network latency is terrible for some businesses and worsens as customers grow increasingly intolerant of slow response times. You must offer an exceptional digital experience to attract and retain internet customers. In a competitive

---

32 www.submarinecablemap.com (n.d.) *Submarine cable map.* [online] Available at: https://www.submarinecablemap.com/.

33 Law Insider (n.d.). *Cable landing station definition.* [online] Available at: https://www.lawinsider.com/dictionary/cable-landing-station [Accessed 19 Aug. 2022].

34 Editor (2019). *Point of Presence (POP).* [online] Network Encyclopaedia. Available at: https://networkencyclopedia.com/point-of-presence-pop/ [Accessed 19 Aug. 2022].

35 A cross connect is a point-to-point cable link between two customers in the data centre or business exchange. With cross connects you receive a fast, convenient, and affordable integration with business partners and service providers within the data centre ecosystem. You also get highly reliable, extremely low-latency communication, system integration, and data exchange.

36 Unbounce (n.d.). *Think fast: The page speed report stats & trends for marketers.* [online] Available at: https://unbounce.com/page-speed-report/.

business climate, organisations struggle to maintain loyalty and engage users online. Consumers' loyalty is transient with low switching costs, and user expectations for how digital experiences should perform have never been higher.

Data centres produce a significant amount of network traffic. The ideal situation is to have multiple service providers that use diverse paths and separate[37] main points of entry (MPOE). The availability of fibre at a potential data centre location is paramount. Data centres require dependable, robust, and scalable network connections. These network connection needs should be considered at the beginning of the planning process. The bandwidth and latency needs of each data centre may vary depending on the type of traffic that traverses the network. For example, financial applications are likely to be more sensitive to latency issues than website hosting. Capacity and performance demands are also expected to change due to economic conditions, application availability needs, and target markets. Recognising the type of data traffic that will pass over the network is essential.

The latency issue may or may not be pivotal in the decision of where you locate your data centre. Technical due diligence of the fibre network adjacent to the prospective site is recommended, and many firms are[38] available that do desktop studies and virtual network path analysis.[39]

Geopolitical Risks, Laws, and Regulations

Geopolitical risks identify the socio-political environment considering where a data centre may be located. These socioeconomic, workforce and governmental criteria are all associated with the social and economic stability of the region, the availability of construction and sustaining workforce. Existing regulations, taxation, and incentives are all considerations.

For example, the COVID-19 pandemic has exposed supply chain vulnerabilities and heightened reliance on technology at a macro level. Strategic rivalry between the United States and China was driving global fragmentation as both countries are focused on reducing vulnerabilities and a managed decoupling of their tech sectors.[40]

---

37 For example, a North side and South side entry to the Data Centre building or campus.

38 Analysys Mason (2019). *Home.* [online] Available at: https://www.analysysmason.com/ [Accessed 18 Aug. 2022].

39 www.solarwinds.com (n.d.). *NetPath – easy visual network path analysis | SolarWinds.* [online] Available at: https://www.solarwinds.com/network-performance-monitor/use-cases/ netpath [Accessed 19 Aug. 2022].

40 O'hara, K. and Hall, W. (2021). *Four Internets: Data, Geopolitics, and the Governance of Cyberspace.* New York, NY: Oxford University Press.

As with any investment, it is preferred to select a safe country where the government is stable, political tensions remain within the bounds of the democratic arena, and the use of violence is uncommon, if not non-existent. A ranking developed annually by the World Bank[41] is an often used and excellent source to guide these types of considerations.

Environmental and supply-side constraints may prompt a change in policy within a jurisdiction. A recent example would be the network power constraints[42] experience within Ireland and Netherlands.[43] In fact, the data centre industry in Europe has recognised[44] that the enormous power consumption poses a challenge to a sustainable and secure power system and prompts a review by the data centre operators and trade associations. Whilst the data centre industry may influence the European Union policies. The current EN50600 guidance standard[45] may seem likely to progress in law to enable the twin transitions of digitalisation and decarbonisation of our economy and society. How these pending constraints and policies impact each country and jurisdiction is unclear in a changing and dynamic response to global warming and the transition targets set to net zero emissions.

It is, therefore, essential to consider current policy and legislative constraints as part of a site selection process in the target location.

Whether you want an environmentally controlled facility that is highly secure but also low profile and possibly hidden in a remote locale is just one of several factors considered by data centre developers when choosing a site.

Availability and Cost of Electrical Power

A functional data centre may not be possible without the appropriate available power from a network supply. Whilst on-site power generation may be likely by

41 The World Bank (2018). *Countries | Data.* [online] Worldbank.org. Available at: https://data.worldbank.org/country.

42 Government of Ireland Statement on the Role of Data Centres in Ireland's Enterprise Strategy, July 2022.

43 Netherlands National Strategy on Planning (NOVI) is broadly based on the Spatial Strategy Data Centres 2030 (*Ruimtelijke Strategie Data Centres*); the policy was imbedded in the policy decision number 2.6 of the NOVI, which specifically relates to data centres.

44 Anon (n.d.). *Climate neutral data centre pact – the green deal need green infrastructure.* [online] Available at: https://www.climateneutraldatacentre.net/.

45 The European Standard for Data Centre Infrastructure EN50600 (n.d.). [online] Available at: https://www.en-standard.eu/csn-en-50600-3-1-information-technology-data-centre-facilities-and-infrastructures-part-3-1-management-and-operational-information/.

the combustion of either fossil fuel(s) or wind energy, the optimal resilience[46] of the power systems may be compromised if no network supply is available.

The operation of a data centre is generally for every hour of the day with minimal, if any, downtime. Therefore, the cost of electrical power[47] is crucial when selecting a site location, as the operational costs may be inputted into the financial modelling. Denmark has the highest electricity prices worldwide.[48] In Poland, users pay about half as much, while in Sweden, consumers spend approximately one-third of that amount due to the increased availability of hydroelectric power generation. The USA prices are similar to those in the Nordics. When investors and operators consider comparable locations, the ratio of OPEX to CAPEX and return on investment analysis will be dominated by the energy costs; for example, a slight difference of, say, $1 per Kilowatt-hour may result[49] in an $8,760,000.00 difference per MW in a year.

Natural Resources

A connectable substantial water resource will permit adiabatic cooling[50] options to be deployed. The availability of natural gas may allow the data centre to be powered by gas engines that can provide operators with a reduced carbon footprint, increased profitability due to the gas engine asset's flexibility and improved supply security by addressing grid capacity limitations.

Airport Flight Paths

Locating a data centre outside the airport flight paths is recommended, particularly outside any holding pattern where high air traffic density may occur and the altitudes are relatively low. If an airport is near a potential data centre site location, be mindful of the flight paths that incoming and outgoing planes frequently

---

46 See Chapter 5 of this book for power solutions.

47 GlobalPetrpPrices.com (n.d.). *Gasoline and diesel prices by country | GlobalPetrolPrices.com.* [online] Available at: https://www.globalpetrolprices.com.

48 Statista (n.d.). *Electricity prices worldwide 2021.* [online] Available at: https://www.statista.com/statistics/263492/electricity-prices-in-selected-countries/#:~:text=Global%20household%20electricity%20prices%202021%2C%20by%20select%20country&text=Denmark%20and%20Germany%20had%20one [Accessed 16 Aug. 2022].

49 Based on a 24-hour 365 day operation.

50 See Chapter 5 of this book for cooling solutions.

follow. Whilst plane crashes or debris falling from aircraft are unusual, the effect can be devastating if something does impact your data centre.

The local aviation authority can inform you of the flight path of any airport. However, it is also preferred if a data centre has reasonable access to and from an airport and indicates the educated populace may be close.

Electromagnetic Interference

Electromagnetic field interference (EMI)[51] disturbs the normal functioning of an electronic device and is highly harmful to IT equipment and the data centre environment and may result in unexplained data errors. EMI at low frequency caused by power sources may corrupt server data or erases entire hard drives.

In many cases, EMI causes significant threats to information security. Many information technology[52] types of equipment see malicious attacks on their immunity due to high-power EMI. Malfunctions and system crashes are also two significant threats to information security.

If your site selection is potentially near an identified source of interference, locate the data centre as far away as possible from that source to limit the effects. Significant sources of interference, such as telecommunication signal facilities, airports, or electrical railways, may interfere with the data centre servers and networking devices if they are nearby. These potential interferences are regulated[53] in the European Union and other jurisdictions. So it is advisable to consider the regulatory environment when locating your data centre.

All shielding products, such as coatings, compounds, and metals,[54] block electromagnetic interference. However, many of them are meant for use on individual devices rather than over a large data centre. The best protection is the distance from the source of interference. That is because electromagnetic interference corresponds to the inverse square law of physics, which asserts that a quantity of impact is inversely proportional to the square of the distance from a source point. This law applies to gravity, electric fields, light, sound, and radiation. In effect, if a

---

51 Pickard, T. and And, E. (2020). *Guide to Cables and Cable Management.* London: The Institution Of Engineering And Technology.

52 IT meaning Information Technology.

53 Information technology. Data centre facilities and infrastructures Guidance to the application of EN 50600 series. (n.d.).

54 www.essentracomponents.com (n.d.). *EMI and shielding data centres and enclosure | Global Manufacturer & Distributor of Component Solutions — Essentra Components.* [online] Available at: https://www.essentracomponents.com/en-gb/news/product-resources/emi-and-shielding-data-centres-and-enclosures [Accessed 19 Aug. 2022].

data centre is positioned twice as far from a source of electromagnetic interference, it receives only 1/4th of the radiation. Similarly, if a data centre is ten times as far away, it receives only 1/100th.

A potential site with overhead power lines may need to be diverted or undergrounded to reduce the risk of EMI. In any event, the power network provider will have regulations regarding the distance between power cables and buildings.

Typically shield enclosures, power and signal line filters, and RF-tight doors are all methods of protecting the data centre but are likely expensive. For a greenfield project, such protection may result in a 5% or more cost increase.

Taxes, Regulations, and Incentives

Data centres generate revenues, skilled jobs and power usage that make them attractive to government entities and utilities. The local, regional, and national taxes, regulations and incentives may affect virtually every consideration for choosing a data centre location.

Taxes may be levied at the local, regional, and national governmental entity levels. It may be helpful to consider whether data centre equipment is taxed as property and whether sales tax is applied to construction materials, data centre equipment and ongoing data centre supplies. In general terms, it may be found that the lower the taxation rate, the lower the data centre cost.

Similar to taxes, **Regulations** may be enforced at the local, regional, and national levels. Considering the potential jurisdiction of a data centre may be vital. Land acquisition and title may be subject to engagement with the statutory authority having jurisdiction. Many developers will take options[55] on land and concurrently develop an understanding of the current zoning and the appropriate process for change. In most Tier One Countries,[56] planning regulations are complex and highly prescriptive. In the European Union, an environmental impact assessment (EIA) is generally required for data centre-type developments. In most cases, environmental impact statement (EIS) is mandatory as part of the planning permission process. A natural impact statement (NIS) may also be required, depending on the circumstances of the application. These constraints on the location of a data centre need future consideration and are specified

---

55 With option to purchase agreements (also known as a lease option), the buyer is given the right to buy the land for a certain period of time, and it may be subject to certain trigger events. If a specified event occurs, the buyer has an absolute right to purchase the land.

56 blogs.worldbank.org (n.d.). *New World Bank country classifications by income level: 2022–2023.* [online] Available at: https://blogs.worldbank.org/opendata/new-world-bank-country-classifications-income-level-2022-2023.

within each jurisdiction. It may be worth noting that reputational damage and delay may be pretty significant when things do not go to plan. For example, Apple, having procured land for circa $15m in 2014, applied for planning permission to construct a data centre in Athenry, County Galway, Ireland, in 2015. Apple later abandoned its plans in 2018 after progress on the project stalled in the face of various legal actions and local opposition to the project. Though its plans to build a data centre by 2017 got approved by the local council, the appeals process kept bringing the case back to the Irish Planning Board and the Irish Commercial Courts.[57] If any lessons are realised from the Athenry debacle, planning applications ought to be concisely formed to comply with the jurisdictional constraints.

Incentives

When deciding where to locate a project, those involved in Site selection should pay close attention to the incentive offerings. In the United States, many states offer:[58]

a) Property tax exemptions.
b) Sales tax exemptions on certain of the equipment used in the facility, electricity charges, and construction costs.
c) Complete tax abatement on ad valorem and municipal taxes for data centres.

In Europe, the incentives may seem more discrete. Sweden and Scandinavia have become home to some of the world's largest digital companies in recent years. Node Pole, in the north of Sweden, for example, offers a brief time to market with many sites that are available to build with permits already in place and provides concierge help with inward data centre enterprises. All jurisdictions have special incentives. A good starting point to investigate these incentives may be the Arcadis Index.[59]

The incentive landscape, however, may be changing. The climate crisis continues to worsen underneath the sunny picture of rising demand and market

---

57 AppleInsider (n.d.). *Apple revives controversial Athenry, Ireland datacentre plan.* [online] Available at: https://appleinsider.com/articles/21/06/25/apple-revives-controversial-athenry-ireland-datacentre-plan [Accessed 18 Aug. 2022].

58 https://info.siteselectiongroup.com/blog/data-centre-economic-incentive-landscape-in-2019.

59 www.arcadis.com (n.d.). *The arcadis data centre location index 2021.* [online] Available at: https://www.arcadis.com/en/knowledge-hub/perspectives/asia/2021/data-centre [Accessed 18 Aug. 2022].

confidence. The regulatory landscape around net zero[60] creates new risks, coupled with growing client and consumer expectations. There may well be two limbs of incentive, both penalties and tax rewards, to prompt data centre performance within net zero.

Businesses may qualify for and be entitled to tax and sustainability incentives. Vital utility service is integral to the operation of your data centre site. The requirement for more power and utilities can significantly impact the data centre site selection decision.

For example, in Denmark, the data centres have great technical potential in collecting and utilising the excess heat for district heating systems, among other things. In the past, the data centres were subject to a tax on the remuneration for this extra heat, reducing the financial incentive for district heating plants to use the excess heat[61] from the data centres. The June 2020 Climate Agreement for Energy and Industry[62] changed this so that from 2021 onwards, there will no longer be a tax on electricity-based surplus heat from data centres. This promotes the possibility of utilising excess heat from data centres.

New trends indicate that data centres require more power than ever, perhaps 200–300 W per square foot. To make the correct site selection, you need to understand the accessible transmission infrastructure to allow the power to the site and whether your prospective utility company will offer a connection and upgrade if necessary. The power connections offered by utility companies operate within a highly regulated environment, often co-dependencies with the planning permission process.

Recently, some utility companies[63] have begun partnering with businesses to bring data centres to fruition. The business benefit of having your data centre located within their service area, the more sophisticated utilities might provide services including researching data centre sites and market demographics. These power utility companies may provide introductions to regional economic development experts, provide infrastructure details, inform you of tax and incentive opportunities, offer turnkey engineering, construction, procurement,

60 Lexico Dictionaries | English (n.d.). *NET ZERO | Meaning & Definition for UK English | Lexico.com.* [online] Available at: https://www.lexico.com/definition/net_zero [Accessed 18 Aug. 2022].

61 Gunning, M. (2020). *Facebook's hyperscale data centre warms Odense.* [online] Tech at Meta. Available at: https://tech.fb.com/odense-data-centre-2 [Accessed 18 Aug. 2022].

62 IEA (n.d.). *Danish climate agreement for energy and industry etc. 2020 of 22 June 2020 (only EE dimension) – policies.* [online] Available at: https://www.iea.org/policies/12139-danish-climate-agreement-for-energy-and-industry-etc-2020-of-22-june-2020-only-ee-dimension [Accessed 18 Aug. 2022].

63 www.energiagroup.com (n.d.). *Huntstown bioenergy.* [online] Available at: https://www.energiagroup.com/renewables/huntstown-bioenergy/ [Accessed 18 Aug. 2022].

and maintenance, and assist with site visits. Like any business enterprise, key business and government partnerships often increase efficiencies, reduce costs, and streamline approvals.

As the data centre generates much heat from all equipment in the plant, this heat may be reused instead of just sending it out in the air. Recent developments[64] in the Nordics heat recovery mean that the residual heat generated in the data centre is donated to the local district heating network that serves domestic houses. The opportunity for heat recovery may differentiate one site from another as the optics of reducing waste and increasing efficiency becomes more compelling.

Know the Stakeholders

Regional participants and community members can sometimes be overlooked in data centre site selection, but they can be the operator's most important ally. Operators know all the benefits their data centres can bring to the adjacent areas, including tax revenue, high-quality jobs, and enhancements to local infrastructure.

Many regions offer abatements on sales, property, and energy taxes to demonstrate their willingness to host data centres. Operators should respond by displaying a readiness to work with local officials and community groups. But even then, the community often needs to be informed of the benefits.

Expect the Unexpected

Despite the best endeavours to comprehensively investigate a location for seismic activity, vulnerability to natural disasters, talent availability and abundance of bandwidth, and unanticipated circumstances can still derail a project. For example, the sheer complexity of data centre construction is a risk. The project typically involves several vendors, subcontractors, and as many as 50 different disciplines in areas like structural, electrical, HVAC, plumbing, fibre networking, and security. The Uptime Institute reported[65] that individual errors trigger the vast majority of data centre failures and that in the most acute cases, nearly three-quarters of operators believed that better management, processes, or configuration could

---

64 Facebook's hyperscale data centre warms Odense, https://tech.fb.com/engineering/2020/07/odense-data-centre-2/.

65 Uptime Institute Blog (2019). *How to avoid outages: Try harder!* [online] Available at: https://journal.uptimeinstitute.com/how-to-avoid-outages-try-harder/.

have prevented interruption. Skilled and experienced project managers must supervise the effort from start to finish to ensure smooth operation.

Numerous peripheral factors need to be considered, including the proximity of transportation sources. If hazardous materials are transported on nearby railroad lines or by a truck on adjacent highways, the impact of a derailment or crash could be disastrous. The potential of your neighbours' business says a chemical plant down the street may issue noxious fumes that could interfere with your cooling system or create an unpleasant work environment, and local easement rules for a walking or bicycle path that goes inside your property create a security problem.

Traditional Due Diligence

Whilst data centre site selection is unique, establishing site title and ownership is essential. It is good practice to engage a regulated firm[66] to consider matters reasonably apparent from a site inspection. The act of conveyancing ought to identify the use, ownership, or occupation of the land or property, such as boundary issues, shared services, rights of access, common areas, rights of way, etc. Pre-contract due diligence ought to bring these to the client's attention for referral to the client's legal team.

The lawyers may well seek copies of related deeds, surveys, title insurance policies (and all documents referred to therein), title opinions, certificates of occupancy, easements, zoning variances, condemnation or eminent domain orders or proceedings, deeds of trust, mortgages, and fixture lien filings. A data centre site requires the same due diligence as other types of development regarding the land procurement transaction.

Retrofitting Commercial Buildings for Data Centres

Working within the constraints of modern data centres' power, fibre network, and cooling requirements, a purpose-built facility provides the most significant efficiencies, reliability, and room for phased expansion with a modular build. However, retrofitting an existing commercial building can still result in a highly efficient and effective data centre environment and save budget and time for deployment. Naturally, many of the considerations for data centre site selection apply when considering existing buildings for a retrofit.

---

66 RICS professional standards and guidance; global technical due diligence of commercial property.

Once you have located a building with sufficient space and can relatively easily be built out to support power and network requirements, HVAC and power routing are most likely to be two of the highest costs as they require specialist contractors and are essential to data centre functionality. Consideration may need to be given to installing supplementary or different transformers on the site. Ductwork may need to be rerouted according to new layouts. The void space between the floor and ceiling must also support systems like HVAC and power.

Regulations concerning zoning and permits must also be considered. Although construction permits might be easier to obtain for a retrofit project, there may be constraints on what equipment is placed outside. The faster time to market presented by a retrofit in comparison to a new build project may depend on how fast those permits can be obtained.

Raised floors are typically seen in older type buildings where power and data cables occupied the floor void that was used for another consideration. Modern data centre designs may not require a raised floor and may use a top entry for power and fibre connection. Structural surveys will determine if the roof and floor will support the weight of the equipment that will be installed.

Site security and fire suppression systems may need to be upgraded to be suitable for data centre use. You may need to improve building security measures with fencing, CCTV systems, or hardening against earthquakes, floods, or other disasters.

Eventually, a very attractive site may lose out to an initially less attractive location due to a mixture of the above factors. Developers should rank these aspects by their significance and then appropriately apply that scale to sites. This will help identify immediate deal breakers and save the time and expense of engaging with expert consultants on inferior sites.

Clusters

In theory, a data centre can be built anywhere with power and connectivity, but location impacts the quality of service the facility can provide to its customers.

The effect of clusters and peering points cannot be underestimated. An economy of scale comes with new data centres joining an established cluster and peering points,[67] namely the ability to 'plug in' to an already established infrastructure.

---

67 Peering involves two networks coming together to exchange traffic with each other freely and for mutual benefit. This 'mutual benefit' is most often the motivation behind peering, which is often described solely by 'reduced costs for transit services'.

The historical trend has been when telecom companies built the first facilities and established footholds with the later emergence of hyperscale and colocation operators in data centre markets. Increasingly data centres are popping up next to one another and forming clusters. An extension of the digital ecosystem has been the development of 'clusters' as geographic concentrations of interconnected entities. Clustering has many benefits, including increasing innovation, collaboration, knowledge sharing, and other spillovers which enhance productivity and competitiveness, e.g. Silicon Valley as a technology cluster. These benefits are derived from a concentrated base of potential customers, suppliers, and resources, including skilled labour. A key attraction for companies in data-intensive industries is to be close to both their own data and those of suppliers and customers to increase efficiency in Business-to-business and business-to-consumer transactions.

The clustering effect is evident in every key data centre location across the globe, including London, Frankfurt, Amsterdam and Loudoun County, Virginia, and Dublin[68] is another example of this.

It is understood that the Northern Virginia data centre community began in the 1990s when key players such as AOL[69] and Equinix added facilities in the area. But Northern Virginia's reputation as a prime location for data centres and colocation was ultimately solidified when MAE-East,[70] one of the first large internet peering exchanges, was relocated there in 1998. The relocation of MAE-East meant that most of the world's internet traffic flowed through Northern Virginia.[71]

Subsea fibre landing stations are also setting a cluster trend. It is worth noting that where submarine cables land into the submarine line termination equipment (SLTE) extends all the way into metro-area colocation data centres. Cables over the past decade have landed directly in colocation facilities or carrier hotels in London, New York, and Hong Kong. Some titans of colocation, like Equinix and others, are landing deals to get subsea cables directly into their buildings.

The clustering phenomenon is well understood in the Luleå experience. Luleå is a coastal town in northern Sweden, less than 100 miles from the Arctic circle, with a population just shy of 50,000, mainly known for medieval wooden houses and natural resources.

---

68 Mdmeng.ie (2013). *Project – T-50 telecoms network* [online] Available at: http://www.mdmeng.ie/t-50_telecoms_network.php [Accessed 2 Sep. 2022].

69 dbpedia.org (n.d.). *About: AOL.* [online] Available at: https://dbpedia.org/page/AOL [Accessed 2 Sep. 2022].

70 Wikipedia (2022). *MAE-East.* [online] Available at: https://en.wikipedia.org/wiki/MAE-East [Accessed 3 Sep. 2022].

71 Lightyear (n.d.). *How Ashburn, VA became the Colocation Mecca known as Data Centre Alley.* [online] Available at: https://lightyear.ai/blogs/ashburn-colocation-data-centre-alley.

Some 10 years ago, it was down on its luck. Luleå's most significant industry was reeling from a last-minute decision to pull a new steel plant, with roads built, ground prepared, and thousands of new jobs promised. Now the town watched, powerless, while the fruits of the budding Luleå Technical University were steadily cannibalised from abroad,[72] each new promising academic spin-off immediately snapped by a major US tech firm and relocated to London, Berlin, or Silicon Valley.

Then, in 2011, Facebook came calling. The world's most popular website sought a spot for its first hyperscale data centre beyond US soil. The colossal server farm would consolidate the billions of photographs, messages, and likes into a single storage and processing behemoth, where massive computing and power consumption efficiencies would help attack the company's spiralling electricity bill. In Luleå, it liked what it saw, and later this prompted a cluster effect in and around Luleå of new data centres, thanks in part to the region's surfeit of cheap power and solid digital connectivity.

However, more recently, many hyperscalers[73] that dominate cloud, network, and internet services can enter a new or somewhat immature market and simply begin a significant build. This change has led to rapid increases in market size, particularly in cities across Southeast Asia, South America, and soon, sub-Saharan Africa. It is anticipated that secondary markets will continue to benefit as specific primary markets restrict power usage and as sustainability demands pressure the industry. Indeed, widespread data centre proliferation has led to concerns in many key global markets. For example, Dublin[74] and Amsterdam[75] are locations with deep data centre capabilities but have placed moratoriums on new data centre development declared in recent years. What has happened in Ireland and the Netherlands is an example of what may happen in other European markets. Sustainability is a broader European topic. Data centre growth is a given, and

72 ww3.rics.org (n.d.). *Data centres: Smart solution or eco nightmare?* [online] Available at: https://ww3.rics.org/uk/en/modus/technology-and-data/harnessing-data/how-the-cloud-created-the-perfect-storm.html [Accessed 4 Sep. 2022].

73 Leopold, G. (2017). *Hyperscalers emerging from 'hype phase'.* [online] HPC wire. Available at: https://www.hpcwire.com/2017/04/12/hyperscalers-emerging-hype-phase/ [Accessed 3 Sep. 2022].

74 An Coimisiún um Rialáil Fóntais Commission for Regulation of Utilities Decision Paper An Coimisiún um Rialáil Fóntais Commission for Regulation of Utilities CRU Direction to the System Operators Related to Data Centre Grid Connection Processing Decision (2021). [online] Available at: https://www.cru.ie/wp-content/uploads/2021/11/CRU21124-CRU-Direction-to-the-System-Operators-related-to-Data-Centre-grid-connection-processing.pdf.

75 Comment, P.J. (n.d.). *Amsterdam resumes data centre building, after a year's moratorium.* [online] www.datacentredynamics.com. Available at: https://www.datacentredynamics.com/en/news/amsterdam-resumes-data-centre-building-after-years-moratorium/ [Accessed 3 Sep. 2022].

more local governments will grapple with how to manage growth and reach their sustainability goals. That may be why in January 2021, the European Data Centre Association[76] announced the 'Climate Neutral Data Centre Pact' to make data centres climate-neutral by 2030, and these matters are addressed in Chapter 9 on sustainability.

Companies tend to cluster or colocate other corporate functions around their data centre locations. Clustering delivers economic benefits that have no direct link to the data centre. Data centres drive significant investment in local communications infrastructure, which draws in other businesses. Without the catalyst of the data centre, the network infrastructure would not be upgraded.

The availability of skilled labour in the region to construct and operate a data centre at a reasonable cost is required. In the construction phase, it will be necessary to source concrete workers, steel framers, electricians, and pipe fitters. These trades may be local or migrant workers that have the capacity to work within the jurisdiction of the site.

In the operational phase, the IT personnel, such as network staff, system administrators, and data centre facility managers,[77] will be required to keep the facility running. It is often seen that locating large data centres near a university or technical college provides a readily available source of talent.

Data centres tend to be built anywhere with power and connectivity in clusters. However, location has an influence on the quality of service that the facility is able to provide to its customers.

Connectivity, for example, is a cooperative venture which relies on proximity. Excellent connectivity depends on numerous redundant fibre connections provided by significant bandwidth providers. The preferred way to provide stable and reliable bandwidth at the volumes required by an enterprise-grade data centre is to build connections to different network providers.

Data centres tend to cluster together at significant peering points and are located in close geographic proximity to internet exchanges[78] or peering points, with the benefit of low latency and multiple redundant bandwidth.

---

76 www.iso.org (n.d.). *EUDCA – European Data Centre Association* [online] Available at: https://www.iso.org/organization/4611823.html [Accessed 3 Sep. 2022].

77 Facility managers typically engage contractors to provide maintenance to chillers, UPS power systems, and security.

78 Internet exchange points (IXes or IXPs) are common grounds of IP networking, allowing participant internet service providers (ISPs) to exchange data destined for their respective networks. IXPs are generally located at places with pre-existing connections to multiple distinct networks, i.e. data centres, and operate physical infrastructure (switches) to connect their participants

There can be many reasons for clustering data centres. In a multi-data centres cluster, the replication is per data centre. For example, if you have a data centre in Amsterdam and another in Dublin, then you can control the number of replicas per data centre. This allows the redundancy to tolerate failures in one location and still be operational, serves requests from a location near the user to provide better responsiveness, and balances the load over many server locations.

A single point of failure frequently does cause considerable disruption to operations and consequently leads to revenue loss. In Texas, in February 2021, the record-setting winter storm and subsequent power outage proved to be a reality check for data centres in the state. Although there were no large-scale failures, there were significant issues with electrical failover systems. This is why you need a pervasive risk management plan and policy that apply to the whole organisation.

There is also an economic incentive to clustering. When one data centre moves into a location, they tend to do a lot of work upfront – everything from negotiating tax incentives to building much-needed resources like fibre/power networks. Several times, after this anchor has been created, other data centres will come into that location as most of the enabling work has been done for them. For example, there are tax advantages once real estate investment trusts[79] (REITs) are created and lower costs when power and communications infrastructure are in place already. When the fundamentals have been established, data centre operators tend to swarm.

Irrespective of how much bandwidth a data centre has access to, customers are unavoidably constrained by the physics, and data takes time to travel to the internet infrastructure. Round-trip distances are double the geographic distance because both the request and the response have to traverse that distance which takes time.[80] This time of response is essential to some businesses; surveys[81] constantly show that internet users are quick to give up sites with slow page load times because people want access to data instantaneously, or they may be tolerant to the time taken for a transaction; however, some applications have a high tolerance to high latency, and it all really does depend on the applications.

---

79 Chen, J. (2020). *Owning property via a real estate investment trust.* [online] Investopedia. Available at: https://www.investopedia.com/terms/r/reit.asp.

80 Gartner (n.d.). *Definition of RTT (Round-trip Time) – gartner information technology glossary.* [online] Available at: https://www.gartner.com/en/information-technology/glossary/rtt-round-trip-time [Accessed 21 Aug. 2022]. "RTT (Round-trip Time) Measure (in milliseconds) the latency of a network – that is, the time between initiating a network request and receiving a response. High latency tends to have a more significant impact than bandwidth on the end-user experience in interactive applications, such as Web browsing".

81 Digital.com (n.d.). *1 in 2 visitors abandon a website that takes more than 6 seconds to load.* [online] Available at: https://digital.com/1-in-2-visitors-abandon-a-website-that-takes-more-than-6-seconds-to-load/.

Qualitative Analysis

The decision to locate a data centre will want to consider the balance between cost and risk in selecting a location. The ideal quadrant here is favoured when making this compromise. Risk mitigation also plays a decisive role in pricing. The extent to which providers must implement special building techniques and operating technologies to protect the facility will affect the price. When selecting a data centre location, it may be essential to consider the certification level based on regulatory and compliance requirements in the industry.

Quantitative Analysis

Site selection has two essential characteristics: The first is to identify the target market and the final consumers; the second is that site selection is the first condition for its success. The traditional site selection method is based on qualitative analysis based on the following criteria, as shown in Figure 3.3.

Site selection is a critical process that can significantly impact how quickly critical data centre infrastructure is available. Getting a site ready to build on can take anywhere from a few months to multiple years. Choosing a site that is not zoned correctly requires remediation or will be met with protest by local officials or residents will only add to the time it takes to prepare a site for construction. And time to market can be critical to meeting today's demand.

Once you know which market is your desired data centre destination, it is essential to take the time to evaluate and analyse the sites that are available in that

Qualitative Analysis
Set the fundamental requirements that the data centre shall meet.
Find out the site which meets the requirements in a specific area.
Determine the key factors that may affect the success or failure of the operation.
Analyse the critical characteristics of these respective sites.
Further select several key sites according to the analysis results.
Determine the key factors that may affect the success or failure of the operation.
Write the investigation report on the selection of each site.
Decide on the site selection after comparing the investigation results.

Figure 3.3 Qualitative analysis of site selection.

region. Decision-makers must work closely with their agents and service providers to identify sites, but it is equally important that they get to the sites to see them for themselves. Driving around an area, seeing the sites in person, and looking for those red flags in advance can make all the difference between choosing a site that can be built quickly and meet all of a company's needs and choosing one that will become an expensive, time-consuming project. By investing time at the frontend to make prudent decisions with an eye towards the future, operators can ensure that they have the support of all stakeholders in the project's ongoing success.

4

IT Operations and the Evolution of the Data Centre

Why does a data centre exist? First and foremost, it exists because it's a necessary part of the infrastructure that powers the digital economy and the Information Technology (IT) of many small, medium, and large enterprises. It is part of an ecosystem of high-speed connectivity, the highways connecting data centres with the internet. Some could be compared to logistics hubs (focused on routing and exchange), others to harbors (focus on the landing of sea cables, pre-processing, and re-routing), others are the local post office (focus on last-mile delivery), and yet others could be considered processing or storage facilities. All have in common that they are connected to the same highways (high-speed fibre) to transport their goods – digital resources.

The data centre facility exists as a necessary part, a component, of IT infrastructure. It is in itself, the facility is neither a product nor a complete infrastructure proposition. Without IT equipment, a data centre facility would be like a railway without trains. Despite different business and ownership models ranging from colocation, cloud, and hosting to enterprise and government-owned facilities, the facilities themselves do not change very much. Like a railway, one can run many kinds of trains and models of how to pay for the train on the same track, without the railway changing.

Building on Figure 4.1, let us quickly define the critical pieces without which digital infrastructure, of which data centre facilities are a part, won't function:

- Highways: Fibre-optical networks
- Harbors and exchanges: Fibre-optical networks and network routing equipment
- Processing and storage facilities: Servers and storage systems, with network switching and routing equipment (IT equipment)
- Last-mile delivery centres: Servers and storage systems, with network switching and routing equipment (IT equipment)

Data Centre Essentials: Design, Construction, and Operation of Data Centres for the Non-expert, First Edition. Vincent Fogarty and Sophia Flucker.

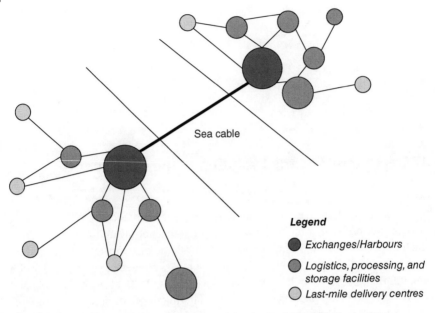

Sea cable

Legend

● Exchanges/Harbours

● Logistics, processing, and storage facilities

○ Last-mile delivery centres

Figure 4.1 Illustration of data centre composition, similarity to logistics, and the highways connecting the facilities.

It's worth noting that all of these critical pieces use electrical power to turn electrons into digital or optical signals. The output is the digital realm as we know it today and heat energy is the result of the conversion of electrical energy.

Each of these pieces lives inside a building, a data centre facility (short: facility), with its electrical (backup batteries, transformation, and distribution), and mechanical equipment (cooling and backup generators). The facility, therefore, is a **component** of digital infrastructure and not the infrastructure in and of itself. To illustrate the importance of this point, imagine a facility without any IT equipment inside of it. In that case, there is no need for cooling, electrical power usage is a minimum, and the network bandwidth that is connecting the facility to the next exchange is not used (no trucks leaving the facility). In short, a facility, without IT equipment inside it, has no purpose, the same as a distribution centre has no purpose without a shipping and receiving company utilising it.

The way data centre facilities have evolved and created an industry for themselves, it has become the narrative that they have a purpose in themselves, their mere existence creating a digital economy in a country or region. As the examples before have shown, that is not the case. Unlike office space, a data centre facility is purpose-built to host IT or networking equipment. To make it more complex, each facility is built to serve a specific IT architecture paradigm (such as cloud infrastructure, a hosting business, a banking IT architecture, enterprise, etc.). In the past, when a facility is not occupied right after it was built, it runs the risk of quickly becoming outdated again, as IT architecture was fast evolving. This, however, is changing as the IT infrastructure itself is becoming a commodity, and with it, the facility.

What has created this view of a data centre facility as a standalone entity is the commoditisation of its parts, such as electrical distribution, backup generation, and cooling. In the past, dedicated IT Operations staff ('IT Ops' or 'Ops') used to be responsible for running the end-to-end infrastructure – from the facility to the IT equipment and networks. As Ops people became more specialised and evolved more towards the IT side, supporting software engineers to build applications rather than the facility side, the facility became an outsourced commodity. From the perspective of most IT Ops teams, a data centre facility has become something like an office space for servers and networks, with standard expectations on availability ('uptime') and standard designs to guarantee that the availability requirements are met (often expected to operate > 99.99% per year).

Most of these standard requirements and designs have been developed by IT Ops people themselves when they were still building and running the data centre facilities themselves. Through standardisation, the aim was to enable the possibility for others, such as traditional commercial real estate firms, who do not necessarily understand anything about IT, to be able to build and operate data centre facilities, and do so in a way that could meet the high standards and availability expectations of Ops staff. Interestingly, this most often excluded the network part which is still run by IT Ops staff, leaving only the passive (non-IT) infrastructure to the operators of data centre facilities.

This development resulted in what we know now as colocation facilities, one of the most common business models in the IT infrastructure space. These are buildings that meet the standards of the IT sector and can be rented on a rack-basis (think of it as hotel rooms), with electricity, cooling, and internet mostly included. A hotel for servers and networks. It had the side effect of shortening the path between different companies' IT infrastructure, which enabled the short-distance and fast transfer of information and sometimes a better integration of business processes with each other, for example, in the case of banks and their financial trading platforms.

Throughout this, you will notice the one driving force: the requirements as set by IT Ops staff which are based on their architecture design for IT infrastructure, and the software applications running on the IT infrastructure. As data centre facilities are part of IT infrastructure, we should have a view of the big picture to understand the role of the data centre facility and its auxiliary infrastructure.

Beginning of IT Infrastructure

Most people's first experience with IT is with personal computers such as an Atari or Commodore, an Apple II, or depending on your age, maybe with a Dell computer or IBM Thinkpad. Whatever it was, it likely still served as your own IT infrastructure. It had all the applications you were using installed on it and it would perform all the functions you required from it (word processing, spreadsheets, calculations, and storing information) without being connected to the internet or any other network. This computer and its operating system (maybe Microsoft Windows, Mac OS, or a Linux Distribution) represented your IT infrastructure.

This is also how digitalisation started in many businesses, using the computer as a digital typewriter, a calculator, or to collect and store data on it. Yet each employee had a computer, with no centralised storage or database system on which they could work collaboratively.

The real value, however, in digitalisation was exactly that. The idea of being able to connect computers and allowing different people in an organisation to collaboratively work on a database, such as a customer database or a ledger and a product register. This idea was not invented in the enterprise, as with most ideas at the beginning of the computer age, they came from academia, universities, hobbyists, and through military complexes such as (D)ARPA. For more information on the beginning of the computer age, I recommend the book 'Dealers of Lightning: Xerox PARC and the Dawn of the Computer Age').

Let's stick with the example of a shared address book or customer database. When connecting computers to a network to collaboratively work on the database, one computer has to 'host' the database while the others connect to it and submit entries. The host computer is what we know as 'servers' – they **serve** the database that everyone is working on.

And this was the beginning of the IT infrastructure. Now, for the first time, a specific set of computers where required, with the following attributes:

1) to be connected to a network that is strong enough for multiple other computers to connect to it
2) to be secured, as the most critical data of the organisation (e.g. customer database) is on it
3) to be made redundant, in case either data or a server is lost
4) to be highly available because when the server is not on the line ('online')/accessible, the clients cannot access the central database
5) to have more computing power and storage capacity than the computers which are connecting to it, as the database will require a lot more space, and providing access to many clients requires more computational power

In the beginning, these requirements were not properly encoded in standards and norms. It was the IT people (in those days, there was no difference between operational staff, software engineers, and IT support staff) that serviced the clients, who also ran the servers, and likely sat next to them on the desk where they work. In most organisations, there was one or a handful of people doing this job – the beginning of the IT department. If your computer does not work or you cannot connect to the customer database, it was those IT people that provided the IT service.

However, looking at these requirements in today's context, one will quickly recognise the Tier Rating system (https://uptimeinstitute.com/tiers) that is most common among data centres today as an encoded version of these requirements.

Before this evolved into data centres, however, it evolved into mainframe computers, shown on Figure 4.2 which solved the redundancy requirement by creating a self-contained system, as well as server rooms, which contained storage, networking, and computing equipment as separate systems.

Bringing Enterprise IT to Maturity

As more and more engineering and business processes became digitalised – from customer relationships to databases, inventory systems, resource planning, etc. – the infrastructure required to run all of these applications became larger.

It is safe to say that the majority of this digitalisation was process-driven and started with either off-the-shelf applications, such as databases or enterprise resource planning systems (e.g. from Oracle, Sun Microsystems, SAP, Microsoft, etc.) that were tailored to map business processes in a digital context.

All of these applications required the enterprise to invest in IT infrastructure themselves, purchasing server equipment, networking as well as providing the rooms to put everything in what we often refer to as server rooms. Eventually, as IT departments grew and more and more parts of the business became digitalised, the server rooms needed to mature and became dedicated buildings – what we today refer to as enterprise data centres (Figure 4.3).

Figure 4.2 A single-frame IBM z15 mainframe. Larger-capacity models can have up to four total frames. *Source:* Agiorgio/ Wikimedia Commons/CC BY-SA 4.0.

Figure 4.3 Example of a FedEx mainframe data centre. *Source:* SWANSON RINK/ https://baxtel.com/data-center/fedex-enterprise-data-center-west-edc-w/photos/last accessed 07 December 2022.

IT Applications as Standalone Products – The Digital Economy

While enterprises were busy digitalising their business processes, a new form of software and companies emerged. These digital product manufacturers (think Aliweb, MySpace, Yahoo, or Google). They created software-based products that could only be accessed and utilised through the internet, the entire value-creation process went digital. They became the start of the digital economy and grew rapidly through the rise of the internet, first static websites, later mobile applications, to interactive Web applications to digital advertising and user-generated content.

For simple websites (basically text documents), it was possible to run them from a simple home computer (turning it into a server) connected to the internet.

Consider the CERN internet server which was shown in Figure 4.4. It was using a NeXT Cube to host websites for the World Wide Web. The NeXT cube that is shown in Figure 4.4 and its object-oriented development tools and libraries were used by Tim Berners-Lee and Robert Cailliau at CERN to develop the world's first Web server (CERN httpd) and Web browser (World Wide Web).

Figure 4.4 The world's first Web server, a NeXT Computer workstation with Ethernet, 1990. The case label reads: 'This machine is a server. DO NOT POWER IT DOWN!!'.
Source: Coolcaesar Wikimedia Commons/CC BY-SA 3.0.

The internet quickly evolved as a means to host noninteractive websites to store and share information. With no underlying business model (most of the content was freely accessible), it remained difficult for the people who were hosting websites to invest the necessary capital into server and networking equipment or server rooms to expand and serve the ever-growing internet community. Even more importantly, hosting websites with a lot of visitors ('traffic') required a strong network connection to the internet.

At the same time, internet service providers ('ISPs') emerged (many of which were run by one or two IT people). ISPs provided dial-up connections for people to connect to the internet. The ISP owners would connect customers via their existing phone line to a 'server room' (which in some cases was the basement of the ISP owners' houses or rented space in an office building). Within the server room, the ISPs would connect all the phone lines to one bigger internet connection (sometimes that was already fibre-optical or DSL-based lines originally intended for enterprise and university connectivity). Renting these larger connections was expensive, but by aggregating demand, it became a financially feasible business, with profits to invest in further expanding the infrastructure.

These ISPs were the predecessors of internet exchanges or peering points (IXPs) which emerged later to enable ISPs to exchange traffic and connect on a neutral internet exchange. ISPs would later also connect businesses and offices and act as regional demand aggregators. You can think of them as the highway ramps for cars, connecting many regional roads to one high-speed highway system.

As hosting websites required fast internet access, it became a side-business for many of these ISPs to offer 'hosting', which meant allowing the owners of the websites to put the computers that served them ('servers') into their facilities and connect them directly to the high-speed (highway) network.

This is an important development to recognise as it happened in parallel to the development of the enterprise and corporate IT sectors and would only become intertwined at a later stage. Both the business of connecting households and corporate offices to the internet as well as hosting websites for hobbyists to offer content on the internet was the first time the notion of a digital economy emerged.

It created value purely by providing digital content and access to it (examples include Netcom in the United States, or T-Online, Freenet, and World-NET in Europe). Telecommunications companies during that time benefited significantly as they owned the main infrastructure, telephone lines, and copper/fibre cables that were driving the increasing connectivity of the internet.

With the number of people connected to the internet increasing, new digital products emerged – from email services, simple games, gambling/betting platforms, internet chat rooms, and eventually with bandwidth increasing, access to pornography, and trading of illegal copies of music, movies, books, etc. The internet became the wild-west, an unregulated digital space in which the government was neither responsible nor had the power to enforce national laws. It became possible to monetise the content of others (e.g. illegal copies of films and music) through online payments or very simple advertising. This was not targeted advertising as we know it today – it was pop-up advertising for more of the same – pornography or internet access.

At this point, the only thing in common between Enterprise IT (the information systems deployed by traditional businesses to digitalise processes) and the digital economy (information systems to provide static and interactive content, internet access, and services) were company websites. These websites were 'digital business cards' and sometimes 'digital storefronts' providing information to the public about a business and its offerings. Businesses digitalised their processes, but they did not transform the fundamental products they made into digital ones.

Second Iteration of the Internet

The first version of the internet was built as a mono-directional medium ('read-only internet') – one could consume written information, download content, music, and films, and store it on one's computer; however, uploading content to the internet was more difficult. Further evidence of this where the internet connections which emerged after the dial-up era, such as asynchronous digital subscriber line (ADSL), were asynchronous, meaning they provided a lot more speed to download content from the internet rather than to upload content to it.

The first generation of internet users bringing content themselves to the internet was in the form of 'bulletin boards' – online forums in which people could start conversations and attach images and later also short videos. After a while, more websites emerged that allowed users to upload their own content, kickstarting the movement of 'Web 2.0' and user-generated content. Two noteworthy examples of that movement are Flickr for photos (2004) and Wikipedia for knowledge (2001).

Flickr, especially, presented one of the first versions of a purely digital product and thus a digital business. Amateur photographers could upload and share photos that they took. And by doing so, they enabled anyone with an internet connection to access their work (and potentially buy printed copies). You can see the scale of photographers uploading their work to Flickr on Figure 4.5.

Compare Flickr to Kodak and the difference becomes clear. Kodak was digitalising its processes using IT, e.g. to improve the efficiency of production and logistics. Only much later they realised how to build a digital product. Flickr instead came straight to the internet with a digital product and a new business model – they were digital-by-design, something that is today referred to as a 'digital-native' business. Flickr and others were a new type of business, not requiring capital goods to purchase machinery to manufacture a product. All they needed was computers and high-speed access to the internet.

Now, as I will argue further later when we will talk about cloud infrastructure, the servers that Flickr needed to run its business were significant capital goods in

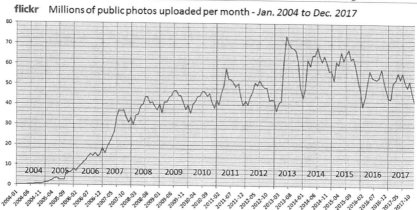

Figure 4.5 Flickr growth of photos per day. *Source:* Franck Michel/Wikimedia Commons/ CC BY 2.0.

the economic sense that required large amounts of money to invest into. High-speed internet access required both capital investment and led to large operating expenses (e.g. for traffic). It's those costs that venture capitalists, in the case of Flickr, Reid Hoffmann, and NFX Capital, financed together with the operating costs of having software engineers and system administrators run the website.

With the 'Web 2.0' came a lot of new digital-native companies, from Facebook to YouTube, to WordPress, Soundcloud, and MySpace. These were companies that hosted user content, most of the time for free, with their added value being that friends (and the public) could see the uploaded content. The business of these companies existed entirely online, meaning it was entirely digital.

Note that this is different, for example, from a company like Amazon, which digitalised an existing physical business (a bookstore) and used a digital application (its website) to create a digital storefront. The core business of Amazon at that point was, however, still physical – logistics, shipping, warehousing, books, etc.

Google, on the flip side, is a prime example of a digital-native business – carving out its niche by making 'the internet searchable'. This was made possible by everything being accessible on the internet without a password, which Google considered 'public information'. And this interpretation enabled Google (and still does) to make a copy of the internet into a large-scale, searchable database to offer its search functionality. In a way, Google pioneered machine-generated insight and value-creation, adding value by storing data and making it searchable, all of which require little human intervention. This is also why most of Google's digital infrastructure today still serves this single purpose: to have a copy of everything on the internet (and with Maps, Mail, Workplace, etc., also a copy of a lot of personal data) in a searchable database.

Companies like Google sell products and services that only exist in the digital realm, which they offer in a global marketplace that is enabled through the global internet. They are truly digital-native companies which means that if the internet were to be turned off, they would have no market and none of their customers would be able to access their products. Likewise, if all their hard drives (data storage systems) were wiped, all program code and data deleted they would have no product.

In the most recent period of the internet, many digital-native businesses selling (digital) consumer products were struggling to find a viable business model. With the struggle came the realisation that the only monetisable commodity these businesses generate is the attention of their users and a relatively accurate profile of their identity as well as preferences. This is visible in the key metric that YouTube is optimising for: time spent watching videos (attention). It is also how Instagram, Facebook, and most other social media companies are measuring their key performance indicators. This attention can be sold to companies as advertising space, in conjunction with predictions on the behaviours of the customers (based on the preference profiles).

In the same period, nondigital-native businesses such as Amazon and eBay digitalised existing businesses that existed in the physical form only earlier, leading to the frenzy of trying to digitalise every existing business and bringing it into the global, internet market (the digital economy). This is a different movement than the emergence of social media and user-generated content. It's essentially what many enterprises failed to recognise as an opportunity when they digitalised their processes. To be fair, however, transforming an existing enterprise into a 'digital native' one is much easier when one can start from scratch and only have to design how the business works for the digital realm. The term 'disrupting' is often used when a traditional business is replaced with a digital counterpart.

'Software is eating the world' – is the most famous quote by Andressen Horowitz from that time, referring to the digitalisation of physical processes, businesses, and things (think of the revolution of TomTom Maps, digitalising the printed atlas or Yelp replacing dictionaries and local magazines). But let's move on to the key ingredients of any digital product.

Key Ingredients for Digital Products

Understanding the history behind digital products and services is important to understand how the underlying infrastructure has evolved and what it means specifically for IT infrastructure, including the data centre facility.

For that, we need to take a closer look at the key ingredients which enabled the explosion of digital product and service companies (misleadingly referred to as 'tech startups', even though most of them do not produce any technology). They are:

1) Access to a global market with no or limited regulation (internet)
2) Access to software developers and IT operations staff
3) Access to financing to fund marketing and operating expenses
4) Access to IT equipment, infrastructure, and network access without capital costs
5) Access to licence-free technology from which a digital product could be made

A **global internet (1)** was given, especially thanks to large, global internet exchanges, most of which evolved from regional ISPs coming together to create national or international exchange platforms – essentially expanding the 'backbone' of the internet. To ensure their neutrality on the traffic, these exchanges are often governed by a nonprofit association, in which all connected parties are equal members (AMS-IX and DE-CIX). Other governance structures exist with academic institutions or government agencies governing internet exchanges.

Telecommunications companies saw a business in connecting people to the internet as they could charge monthly subscriptions (and in the 2000s, still charged end-customers for data traffic as well). Hence, the total addressable market was constantly expanding, at no cost to digital businesses operating on the internet. Many telecommunications companies still lament this fact, recently a renewed call in Europe was made to charge the largest digital companies (such as Netflix, Instagram, or TikTok) to pay a tax for their outsized use of connectivity infrastructure. All these digital companies benefit from telecommunication providers and governments constantly expanding internet access to more people, at no direct additional costs to them.

Access to **talent and people (2)** was not yet as limited as it is today, as many computer hobbyists were eager to turn their hobby into a paying job. Most did not necessarily want to become computer scientists or IT operations staff in a large corporate enterprise. Access to **financing for growth (3)**, at least in some hotspots around the world such as Silicon Valley, was provided by business angels and venture capitalists.

The two key ingredients fueling the explosive growth of digital products, platforms, Web applications, and digital services were **open-source software** and **internet-hosting companies**.

With free and open-source software (FOSS), digital companies had free and unlimited access to the technology needed to quickly build digital products. A good example is WordPress or Drupal, two content management systems that are open-source, and gave rise to millions of digital businesses, blogs, news sites, and other websites without any licence fees being paid to the inventors of WordPress. Another, more fundamental example is all the programming languages that emerged such as Python, PHP, Ruby, JavaScript, and many others, which unlike Java or .NET were not proprietary, and did not require special tooling to use them. Lastly, the technical components that were needed to build digital applications, such as databases, caches, template engines, etc., were all available as open-source software. All that a digital startup company had to do was to either identify a physical business they wanted to digitalise or find a new niche in the user-generated space, e.g. as YouTube did for videos or Twitter for 140-character text snippets. Next, they would quickly glue together open-source technologies to make a product. What looked like a technology company from the outside was merely a business model or product innovation, in the majority of startups, not a technical one.

This brings us to the second component. To launch their business onto the 'digital market', these companies required IT infrastructure – servers, network equipment, fast network access, etc. Buying this infrastructure would have required large amounts of capital investments (not dissimilar from launching a clothing business that requires machinery to turn fabrics into clothing), which these startups did not have access to. The venture capitalists and business angels could not invest millions of dollars into the infrastructure of all these startups which they knew most would fail anyhow. Luckily, this issue had been solved already, in large part thanks to internet-hosting companies, many of which evolved from the ISPs.

As explained before, at the beginning of the internet, financially viable business models were scarce. And the digital businesses that did make money such as illegal trading websites for copyrighted materials, gambling, and others were not legitimate enough to raise large amounts of capital to buy their infrastructure.

Meanwhile, the hobbyists that had operated and built internet service businesses, selling dial-up subscriptions, as well as some of the physical space in their network rooms, were amassing cash that they did not know how to invest. Having front-row seats to the emerging internet businesses, many of which were friends and other hobbyists, struggling to expand due to the lack of capital for expensive IT equipment, they realised they use their cash to help. They started buying servers and rented them to internet businesses that needed them – essentially adding a leasing or financing business to their existing ISP business.

This had two positive effects on the ISPs. It helped further fill the physical space that was necessary for the connectivity part of the business and it helped diversify away from internet subscriptions. For the internet businesses, it provided infrastructure as a pay-per-use, 'Opex-only' model, which could be scaled in line with their business growth and revenues.

This was the birth of the hosting company, many of which grew to significant sizes and are still strong players today, noteworthy examples in Europe are Hetzner, Plus-Server, Strato and 1&1 in Germany, in the United States, Rackspace, Linode, Godaddy, and Digital Ocean or OVH in France, and Leaseweb in the Netherlands amongst many others. Most of them are still privately owned, highly profitable businesses, and all have migrated their 'basement' server rooms into their own or third-party (colocation) data centre facilities, which are tailored to their business model of renting out servers either whole (what is known as 'dedicated servers') or in pieces ('virtual servers').

These hosting companies essentially took the capital risk, assuming that if an internet startup would fail, they would just rent the server to someone else. They could utilise their existing infrastructure (internet access and data centre facilities) and had little additional capital costs.

This led to a competitive price proposition which made the cost-to-fail very low for internet startups (1 server could often be rented for as little as 50–100 EUR or USD a month). All that a startup needed was an idea, a few hundred euros per month to rent some servers from a hosting company, and a few developers (often the founders themselves). After a few months of developing, they had a digital product ready to sell on the largest interconnected market on this planet: the internet.

Difference Between Enterprise IT Infrastructure and Hosting Infrastructure

Now that we've observed the emergence of the internet or digital-native companies and how open-source software in conjunction with hosting companies enabled explosive growth – we can turn back to look at how Enterprise IT has evolved in parallel.

From the outset, it might appear to be very similar to Web hosting. Both own or rent data centre facilities. Both purchase servers, storage, and networking equipment, and both require access to high-speed networks – internet or intranet.

However, it is important to recognise, especially at the time of Web 1.0 and even still during Web 2.0, these were two vastly different types of IT infrastructure. Enterprise IT was designed for reliability and data security deploying Mainframes and high-end hardware from the likes of IBM, Dell, EMC, Cisco, and HP Enterprise. Web hosting, on the other hand, was designed for cost competitiveness, with some servers being self-made or using suppliers such as Supermicro who designed servers to be compact, and space- and cost-efficient.

OVH made a name for themselves, manufacturing their own, super-efficient, liquid-cooled servers inside their facilities.

https://www.datacenterknowledge.com/archives/2013/09/16/ovh-servers
https://youtu.be/Y47RM9zylFY

Another important difference is the types of networks required by either type of IT infrastructure. For enterprise IT, it was important to have private networks to interconnect various locations securely and put all servers and computers on the same network, to ensure a 'closed system'. For hosting companies on the other side, connectivity to internet exchanges (so-called 'uplinks') was much more important. This makes sense, for the enterprise, the market was physical, connecting warehouses, offices, and storefronts, for the digital-native companies, the market was the internet. Both sought fast, reliable, and secure connectivity to their respective markets and environments.

This can be observed easily when looking at two of the largest players in each of the fields of enterprise colocation (Equinix) and hosting (OVH). OVH highlights on their websites their connections (https://www.ovhcloud.com/en/datacenters-ovhcloud/, 32 Tbps with 38 redundant points-of-presence [POPs]) towards internet exchanges around the world and the different connection speeds at which they can be reached (think of it as 'lanes' on the highway, the more lanes you have to an internet exchange, the more trucks you can send through hauling data packages to your customers through internet ports and exchanges).

Now in the case of Equinix, the proposition is much more around the connected nature of all their facilities in hundreds of locations around the world (highways connecting each facility rather than all directed to the internet). Essentially, all of

Equinix's facilities form a fibre ring (https://www.equinix.com/resources/infopapers/ equinix-company-fact-sheet) – a fabric spanning across all their facilities, making it behave like one, global, private network. Now as an enterprise, if I have offices or factories, warehouses or stores in all of the places Equinix is present, the Equinix platform enables me to outsource my server room to their facility. So rather than having a server room in my office building, I instead rent a dedicated fibre connection to an Equinix data centre from an existing connectivity provider. This means each location is securely connected to the Equinix platform, where the IT infrastructure is located and through which all other locations can also be reached.

To summarise: Hosting companies provided the IT infrastructure for the booming digital economy, focusing on low-cost, pay-per-use infrastructure with an emphasis on internet connectivity. Meanwhile, enterprises were still running their own data centre facilities and server rooms, emphasising reliability, availability, data security, and access for all their offices around the country and employees. Enterprise colocation providers were emerging to provide the physical backbone and infrastructure for enterprise IT, however not taking ownership of the IT equipment itself like hosting companies are doing. When enterprises began outsourcing the actual operation of some of the IT infrastructure and thus making IT operations staff partially redundant, the field of Managed Services emerged.

Commoditisation of IT Infrastructure

Two forces began reshaping IT infrastructure and brought the infrastructure design of enterprise IT and the digital economy closer together: Virtualisation and Enterprise Software-as-a-Service (SaaS).

Virtualisation: Servers and Data Centres as Generators for Digital Resources

The first game-changer came with the widespread adoption of virtualisation. Explained in a nontechnical way, virtualisation enabled IT operations staff to divide one physical server into 2, 10, 20, or 100s virtual ones. The result was virtual servers that were securely isolated from the others and each could be treated like a physical server.

It also enabled the aggregation of hundreds of physical servers into a large pool of resources of computing, storage, and network resources that could be shared and dynamically allocated in the form of virtual servers.

From an operations perspective, it added a layer of abstraction on top of the physical servers that before were managed individually and utilised individually. Now each physical server merely was a resource producer (with different speed and quality attributes) in a larger pool and not tied to a single application or specific usage.

The shift to virtualisation was first pioneered by mainframes. When Intel made the ×86 CPU platform a de facto standard for all servers, virtualisation platforms expanded rapidly as the ×86 chips started to support virtualisation technologies.

In 2001, VMWare launched the first commercial server virtualisation platform, VMWare Virtual Server for the ×86 platform which shaped the described, new paradigm in IT infrastructure. The physical servers became a commodity.

Before the widespread adoption of virtualisation, each server was an individual machine that had individual performance configurations. One machine might be configured as a storage server or a database server (usually with a lot of disks and storage space, a lot of memory, and smaller CPUs), another one as a compute server (with few but fast disks, memory, and very fast CPUs), or as a utility server for things like monitoring (with very little performance). Each server had a purpose or role and its hardware was designed and configured to deliver the optimum performance for the destined purpose. In other words, each server was a specialist.

Yet having so many specialists – purpose-built servers – also led to a lot of waste of computing resources, as most of the specialised functions, the servers provided were not needed 24/7 and when needed, not at 100% utilisation. Consider a typical nightly backup scenario – copying a database to the storage system, which requires the full power of the storage system, but during the day, it is not or only sparsely utilised. Another example is a compute-specialised server which may be used a few times a day to perform simulations for engineers, but which is sitting idle when the engineers leave the office. Considering the very high capital costs of computing equipment, this was a significant waste of resources.

Virtualisation promised to solve this problem, first by dividing a server into logical parts, making it possible to run multiple different types of applications on the same server, utilising the available resources in a way that would increase the overall utilisation. In the second generation of virtualisation platforms (e.g. VMWare vSphere, OpenVZ, Plesk/cPanel, later also OpenStack), it became possible to aggregate all servers into one large virtual 'super-server', pooling all the available digital resources (CPU time, memory, storage, and network) together and then enabling the dynamic allocation and consumption of those resources by servers. This was further accelerated by the standardisation of the server technology itself, almost all servers run on the ×86 standard.

Virtualisation also enabled a model that is still very common among any type of hosting company: Oversubscription. When a virtual machine was allocated and sold to a customer, it represented a reservation on digital resources (e.g. 2 vCPUs, 4 GB of memory, and 10 GB of disk space); however, most applications did not use these resources fully, only in certain periods or with predictable variance. This enabled hosting companies to oversubscribe the physical servers, selling more reservations on resources than the server had. Virtualisation platforms handled the reallocation of digital resources from a virtual machine that was idling to one that required more digital resources. This gave an enormous boost to the profitability of the business of hosting companies – the same capital good could be monetised multiple times.

For enterprise IT, it enabled something different: The transition away from mainframes, basically aggregated and often already virtualised 'mega computers' to using commodity servers, such as the ones used by hosting companies. Using virtualisation platforms, they could aggregate those commodity servers now themselves, into their own 'large-scale mainframe' at a fraction of the costs. It also enabled them to leave the proprietary ecosystems of mainframes and the locked-in nature of the software around it. Virtual machines could run any operating system, even different ones on the same physical server, as long as the underlying CPU platform was based on the ×86 standard.

For the first time, Enterprise IT infrastructure started to look a lot more like Web hosting infrastructure, with Web hosting companies having professionalised (offering enterprise-grade availability and security, etc.) and enterprise IT having embraced commodity servers. Albeit, for some time, they would still buy from different original equipment manufacturers (OEMs) and vendors, a clear path to the commoditisation of IT infrastructure became visible on the horizon.

For the people, the IT operations staff, it also created an opportunity to evolve their craft away from the physical work of installing, wiring, and configuring individual servers to working on the virtualised level, getting closer to the applications and their developers. To put it visually: less work in the basement or a server room, more work in the office. After all, in a fully virtualised infrastructure, one only has to connect a server and connect it to the 'fleet' of other machines to participate in the resource pool.

Lastly, virtualisation enabled fault-tolerance on the software layer, e.g. having storage disks automatically replicated across many servers to tolerate the failure of one or having applications migrate from one physical machine to another without any human intervention, improving failure tolerance of physical machines. IT operations staff and software developers started to move a lot of the physical redundancy that used to be carried by the server room infrastructure and data centre infrastructure (e.g. backup power, redundant power feeds, redundant cooling, etc.) into the virtualisation layer.

In the era of virtualisation, a pool of heterogeneous physical servers could now be pooled together to generate digital resources and divide those digital resources into new, virtual servers that could be easily managed, moved, replicated, and scaled. Digital resources could be reallocated instantaneously between applications, new resources could be easily added to the pool. A data centre containing a pool of virtualised servers essentially became a factory for digital resources. A new commodity was born that would become the fuel for the growth of some of the largest digital companies on the planet today, all within the digital economy. More on that later.

Software-as-a-Service: Digital Natives Entering the Enterprise

For some time, there was a clear dividing line between the IT systems and infrastructure of medium- to large-scale traditional businesses and digital-native businesses. In the world of the traditional enterprise, proprietary software was the norm, with an eye on 'business continuity' rather than costs. Video conferencing systems from Cisco and Polycom, database and enterprise resource planning systems from Oracle and SAP, office products and user management from Microsoft, with Visual Basic, C#, .NET and Java, Pascal, etc., being the prevalent programming languages of choice.

On the other hand, the digital-native companies embraced free and open-source software (FOSS): from Postgres and MySQL databases, WordPress and Drupal content management systems, Linux as an operating system, and open-source programming languages such as PHP, Ruby, JavaScript, C++, and Python.

One company that is noteworthy for making the first attempt at bridging these two worlds was Red Hat. Red Hat took open-source software, certified it for enterprise use, and provided service-level agreements as well as support to ensure the continuity and quality of development. They made FOSS enterprise-grade, meeting the strict requirements that until that point was only fulfilled by the enterprise software industry.

Eventually, though, digital startups, discovered that they could play in the field of enterprise software as well, re-packaging free and open-source software with simple user interfaces that would address 'enterprise business challenges' or simply: provide enterprise software that is more enjoyable to use, modern, and intuitive.

This was a strong contrast to the complex user interfaces of Oracle, Microsoft, and SAP, which often required training and large handbooks – the complexity of those systems was thought of as an expression of professionalism and value. Slowly, digital startups began eating away at the enterprise software market, developing new customer relationship management systems (CRMs), workflow automation tools, business process management, database systems, accounting and expense management tools, and so on. Most importantly, they disrupted the traditional licensing model, allowing enterprises to pay only when they a user used the product – pay-per-use, seat- and subscription-based. The era of SaaS was born.

With SaaS, a more important shift in supporting IT infrastructure had happened. Whereas the traditional enterprise software vendors would sell a piece of software that was to be installed on IT infrastructure owned and operated by the enterprise itself, the SaaS products came with their own IT infrastructure. Today, there are very few SaaS products that can be installed on companies' servers, considering many popular tools such as Salesforce, Trello, Asana, Slack, Zoom, Microsoft Teams, Google Docs, Notion, HubSpot – none can run 'on-premise', they only run on their own infrastructure.

The shift to SaaS drove the gradual obsolescence of more and more enterprise IT infrastructure. Within the SaaS companies, the fact that customers were now enterprises increased the availability and security requirements that startups had towards their suppliers of IT infrastructure. Enterprise customers were accustomed to service-level agreements that would guarantee availability, uptime, redundancy, data security, the geographical distribution of backup systems, etc. Complex requirements now had to be passed on to the hosting providers which had previously only cared about costs, rather than meeting enterprise requirements.

It's safe to say, however, that many of the SaaS startups that had to meet enterprise service level requirements, met those using intelligent software, load balancers, or fault-tolerant database systems, rather than asking the physical infrastructure to solve the problem for them. For them, the servers they rented in the data centres were a commodity – a fuel they used to build their software-based redundancy, rather than relying on battery-based backup systems and diesel generators to do the work for them. For software engineers, the solution was always to be found in the software layer.

From the perspective of the enterprise, it started a new era of how companies thought about IT infrastructure that is still shaping the market today. Whereas it was non-negotiable before to have IT infrastructure in-house, it now became more of an open question. If the IT infrastructure was critical to the competitive advantage of the business, it would remain managed by an internal IT operations department, albeit most likely with an outsourced colocation infrastructure to remove the need to invest capital into data centre buildings and infrastructure. If it was not critical, it could be completely outsourced, either to be managed by service providers operating the infrastructure on behalf of the enterprise or be replaced with SaaS-based solutions. Microsoft's Office 365 then did something that was long unthinkable: it moved the access and communication infrastructure – the email server of enterprises, often a sacred piece of infrastructure within the enterprise – to Microsoft's own IT infrastructure, where it would be 'managed' for the enterprise. And if you are willing to outsource your core access and communication infrastructure – why not everything else?

Great Outsourcing of Enterprise IT and the Growth of the Colocation Model

In the beginning, the colocation model was limited to two types of facilities: Existing internet exchange points (IXPs), ISPs, or POPs both on a national scale and regional scale offering to colocate server or networking equipment in their facilities to have better connectivity. The latter is the model of choice for banking and finance, colocating within stock exchanges (or at the next connectivity point) to enable faster trades.

In the beginning, none of this served the enterprise, as most of them were not (internet) connectivity-focused, and had most of their facilities either in their office buildings or near them. When the IT infrastructure grew, many turned to building their own data centres to consolidate the infrastructure, and only a few turned to outsourcing it to someone else. Eventually, enterprise software companies like IBM offered to manage the systems (e.g. often mainframes that IBM sold themselves) on the customer's behalf, either in IBM facilities or at the premises of the customer.

Equally, in the digital economy, the hosting companies that provided most of the IT infrastructure to the many digital companies and startups relied on renting space in office buildings or inside colocation facilities offered by some ISPs. Additionally, they rented space at ISPs and IXPs to provide enhanced connectivity to their IT infrastructure by installing their own high-end networking equipment there. When some of them started to expand and scale, they raised capital to build their own data centre facilities, many of which are still in operation today. However, the majority of hosting companies did not see the physical infrastructure as something that added a lot of value to their core business. Rather they focused on the IT infrastructure itself. They were among the first to start renting space in other data centres from either competitors or telecommunications companies to host their servers.

With the emergence of SaaS and the internet as a global marketplace, many enterprises had to make a shift in their IT infrastructure strategies. Whereas before, the most important part was the interconnection of all offices and access to the IT services and applications of the enterprise, connectivity to the internet became more important. SaaS companies were further driving this trend, as they offered their products only on the internet, and not for installation at the premises and inside the networks of the enterprises.

Further, as more large- and medium-sized enterprises expanded globally, it became too expensive for each company to build and operate their own global networks and connectivity. Instead, they realised, they could run secure and encrypted pseudo-networks on top of the internet. Think of it as using public highways to transport high-value cargo between facilities in armoured vehicles instead of using private roadways. Virtual private networks (VPNs) and not physical private networks quickly became the norm for enterprise connectivity.

This also triggered a shift and further consolidation between the architecture of hosting companies and enterprise IT. Their requirements started to look more similar. Both now required fast access to the internet through internet exchanges and both started to use more and more commodity servers, thanks in large part to virtualisation.

Companies that offered 'neutral connectivity platforms' to customers such as Interxion (now part of Digital Reality), Equinix, Colt, Century Link (now Lumen), and others started to use their data centre facilities – previously mainly used to connect enterprises to each other or to run their very own networks in the case of telcos – as colocation facilities, allowing enterprises to migrate some of their IT infrastructure into theirs. As before they had only networking equipment in the facilities, they now had also the servers and storage systems at the colocation facilities.

This is how colocation (wholesale and retail), as we know it today, was born: Telecommunications and connectivity providers offered their data centres to enterprise and hosting companies (retail and wholesale). Other hosting companies that owned facilities offered space to competitors and enterprises (retail). Everyone was moving closer to the internet and each other.

Further acceleration in digital product areas such as online gaming, video streaming, and global content delivery networks further amplified this new gravity of IT infrastructure toward major internet connectivity. All of the emerging digital technologies demanded lower and lower latencies to provide the best service for their customers.

When Digital Products Scale – or the Invention of Hyperscale

Enterprise IT kept growing and so did hosting companies, but never to a point that they would use the facilities the size of multiple soccer fields as we see them today. The reason is obvious: Neither of them ever had to serve billions of customers across the entire planet.

Think about it: even a large company such as Siemens only serves about 100,000 employees, maybe 1 million people if you include suppliers, with their IT infrastructure. Similarly, hosting companies may host 100 companies' Web presences and applications, which each serve 1000 viewers per month – these are still a very small fraction of what some of the largest digital product companies service today.

We are talking of course about digital-native businesses such as the FAANGs – Facebook, Amazon, Apple, Netflix, and Google. All of them operate the largest portfolios of digital products to date. But there are many more, especially when including the enterprise space, such as Salesforce, Snowflake, Palantir, and many others. To put it in perspective, Netflix has about 208 Million subscribers who spend 3.2 hours per day consuming video content.

It's the consumer-facing digital product companies such as Facebook, Netflix, Google, and YouTube that ran into scaling issues. To deliver their products, they needed global IT infrastructure with the ability to produce digital resources (compute, memory, and storage) at an unprecedented scale and speed. Think about Google for a moment: To deliver its search engine product, it has to have a copy of the entire internet in a searchable database system that can be accessed in sub-milliseconds. And probably with such a valuable database, you need another infrastructure large enough to hold a backup of it.

These companies took the emerging forces – virtualisation and the commoditisation of server hardware – to build not data centres but digital resource factories, housing thousands of commodity servers, virtualised together as one large global digital resource pool. Each facility represents a pool of digital resources – be it storage, computation, or network capacity that can be used to serve its products. These types of facilities became known as Hyperscale data centres, as, unlike colocation facilities, these were designed to only host one large-scale IT infrastructure for one digital-native company. Many of the innovations that we see in IT infrastructure today, such as containerisation and orchestrations, serverless functions as well as large-scale monitoring systems came out of these Hyperscale designs.

When looking at the rare pictures of Hyperscale data centres, the commodity effect of the server hardware is very visible – all servers look the same and looking for OEM logos such as Dell, HPE, and Lenovo is often a waste of time. You can try to find those logos on the Figures 4.6 and 4.7 which give an inside-look into a Google data center.

These commodity machines are made of only essential parts with everything removed that is not essential to the operation of the server itself. Facebook took this commodity movement a step further and created an organisation, the Open Compute Foundation, that maintains and develops open-source hardware designs, further facilitating the transition to commodity servers.

These servers are optimised to remove as many components as possible while at the same time consolidating equipment such as power distribution units and backup batteries on a rack level rather than on an individual server level. Each blade contains only the essential parts of a computer: motherboard, CPUs, networking, memory, and depending on the type of server node – either lots of storage or less. You can see an example of such a server on Figure 4.8.

A Change in Architecture – The Rise of Cloud Infrastructure

The idea of cloud infrastructure and Hyperscale are essentially the same, with the difference that Hyperscale infrastructure was built by digital product companies

Figure 4.6 Google Data Center. *Source:* Google LLC/https://www.google.com/intl/it/about/datacenters/data-security/last accessed 07 December 2022.

Figure 4.7 Google Data Center. *Source:* Google LLC/https://blog.google/inside-google/infrastructure/how-data-center-security-works/last accessed 07 December 2022.

Figure 4.8 Rack with OCP servers. *Source:* Photo by EMPA.

as their own infrastructure – commonly referred to as private cloud in the enterprise world today – whereas cloud infrastructure, as we know it today, is a public marketplace for digital resources, also known as public cloud.

VMWare defines cloud infrastructure as such:

> Cloud computing infrastructure is the collection of hardware and software elements needed to enable cloud computing. It includes computing power, networking, and storage, as well as an interface for users to access their virtualized resources. The virtual resources mirror a physical infrastructure, with components like servers, network switches, memory, and storage clusters.

In 2008, Armbrust et al. defined public and private clouds like this:

> The data centre hardware and software are what we will call a cloud. When a cloud is made available in a pay-as-you-go manner to the general public, we call it a public cloud; the service being sold is utility computing. We use the term private cloud to refer to internal data centres of a business or other organization, not made available to the general public when they are large enough to benefit from the advantages of cloud computing that we discuss here.

In both descriptions, it becomes clear that cloud infrastructure is an abstraction of a data centre facility with the IT infrastructure (servers, storage, and network equipment) that it contains, turning it into a virtual pool of digital resources. The data centre and the servers become one production facility of digital resources, divided into computing-, network- and storage capacity. And these digital resources are commodities, because the means of their production, e.g. the model of the server, and the type of storage disks being used, are becoming invisible and are not used to differentiate the product.

Consider in the past the minuscule differences in servers that IT people were looking for, e.g. what type of memory bus is being used? DDR3 or DDR4 memory, is it ECC buffered (different reliabilities of the hardware being used) or not? How are the disks connected to the motherboard? What network chips are available? None of these attributes is used to differentiate digital resources offered by public cloud providers.

What still differentiate the digital resources provided by cloud infrastructure are three things: the type of CPU (as some CPUs are better at certain tasks than others), what type of disk is attached (directly or via the network), and the available internet and network bandwidth.

Origin of a Market for Digital Resources

Interestingly, it was not the pioneers of building Hyperscale facilities, e.g. Google and Facebook who started selling digital resources through a marketplace. Much of that credit goes to Amazon Web Services (AWS), yet there were many pioneers in the space before. Noteworthy upstarts included not only Linode and Digital Ocean in the United States which offered low-cost virtual machines but also all the traditional hosting companies such as the ones mentioned before in Europe (Hetzner, OVH, Plusserver, 1&1, Strato, Scaleway, Leaseweb and many more) which offered these kinds of virtual resources.

Amazon though took a slightly different approach. At the point when AWS – Amazon's public cloud offering – came along, it was based on a different business model.

Unlike Google and Facebook, Amazon wasn't a digital product company, but rather an e-commerce and logistics company. Its product was not purely digital, but rather a digitalisation of a physical process, with the outcome being the delivery of a physical product to a customer. At its very core, Amazon was about efficiency, and when looking at its own IT infrastructure, it realised it was vastly underutilised. This is the nature of an e-commerce company – customer traffic comes in very predictable peaks (e.g. when people are shopping) and all of the infrastructure must be capable of handling those peaks. This means that in the meantime, the infrastructure is often idle and unused.

So a similar question to the one that ISPs asked themselves in the past arose: What do you do with underutilised infrastructure? And the answer was still the same: You rent it to someone else. So Amazon opened a public marketplace from which people could buy the unused digital resources that the infrastructure was producing but could not use. They offered those digital resources either as on-demand and pay-per-use resources or through an auction system – a spot market.

The spot market is especially interesting as it uses price signals (e.g. AWS is willing to sell 100 virtual resources at $1 each and customers can set a 'strike price' at which they would be willing to buy those resources for an hour or as long as the price is below the set threshold). Unlike the counterparts mentioned before, Amazon did not just sell virtual machines (bundles of digital resources) but rather unbundled them into abstract units that are often referred to as **cloud primitives** or **digital resource primitives**. In Amazon's marketplace, one could buy on-demand resources, reserve resources for 12, 24, or 36 months, or bid on spot price:

- 1 GB of storage with different levels of reliability and redundancy ('storage primitive')
- 1 GB of network traffic ('network primitive')
- A virtual machine with just computation in the form of CPU cycles and memory capacity ('computing primitive')

In particular, the unbundling of computation and storage unlocked a vast new array of product configuration options.

Disconnecting storage from computation alone was a major shift, enabled by massive investments into internal networks. The unbundling of storage and computation is a large part of the reason why public cloud providers can reach such incredibly high utilisation rates (>40–60% vs. 15% in a traditional enterprise data centre). Before this shift, the disk on which the data and the operating system of the virtual machine are stored had to always be within the same server as the network was not as fast (except special fibre-optical connectors such as Infiniband). This created several constraints. First when moving the computation (the virtual machine), one had to move the data as well to a different physical server. Second, if 10 virtual servers each reserved 100 GB of disk space, and the physical server had 1000 GB (1 TB) of storage available, the server was considered full, even if all 10 virtual servers sat idling (meaning they would not consume any computation resources, but block all storage resources of the physical server).

One could say the allocation of virtual servers shifted from storage-driven to computation-driven, which also happens to be the most expensive component of a server (memory and CPUs represent the majority of costs per unit).

As we've seen before, this also enabled the server to be reduced to two types: Computing nodes (servers) with the maximum possible CPUs and memory on them and storage nodes (servers) with the maximum of disks in them. Making them 'the right size' is a balancing act and depends on the network capacity.

Because, as mentioned before, the key to making this work is the network – it has to be fast enough to make the virtual machine behave as if the disk it needs to read/write from is in the same server when in reality it is somewhere else. This can be seen in some of the OCP designs which feature uncommon 2 × 25 Gbit, 4 × 25 Gbit, and even up to 4 × 100 Gbit network link configurations. These links were expensive and uncommon before. This design was what finally enabled the complete commoditisation of server hardware, decoupling storage, computation, and network capacity from a physical machine.

This also meant physical servers could be filled based on the total available computation and would not be constrained by available storage as well. You can see very clearly how this works in the price list of AWS and Microsoft Azure. Each cloud instance or virtual machine type represents a physical generation of computation nodes (servers). They are available in different 'sizes', however, when dividing the available memory (GB) by the hourly price, all of the sizes reveal the same price-per GB of memory. This essentially means that each type of computational node is simply priced on memory, the most expensive component, with CPU prices likely to be factored into the memory costs. It reveals that the limiting factor of the physical machine is no longer storage but memory. This is because memory cannot be sold twice (oversubscribed) and needs to be available when it's reserved (whereas CPU time can vary so greatly that it can easily be oversubscribed as is common practice).

This unbundling of storage and computation has another very big advantage – it allows the computation to shift between different physical machines, e.g. in case a physical server becomes unavailable. As the state of the computation is stored centrally, not on the disk of each server, the computation can be started somewhere else. This works in principle, but due to the secretive nature and how well isolated the digital resources are from the physical infrastructure (and the lack of transparency of public cloud providers), it is impossible to prove or know if this is happening in practice, albeit it is likely. Cloud providers have so far been very secretive about how their digital resource factory works in practice, and likely for good reasons.

Now Amazon realised another important aspect, the rise of digital-native start-ups that would eventually require the same Hyperscale infrastructure that others, such as Facebook and Google already had. A good case in point is Netflix and Dropbox. At the time of writing, Netflix is still using AWS as its primary IT infrastructure provider whereas Dropbox has moved 'out' into their infrastructure, as the unit economics for their storage-based business model are much better on dedicated storage infrastructure.

Amazon offered these companies something that many of the more regional hosting companies could not offer: a global digital resource (IT infrastructure) platform that already served one of the digital behemoths: Amazon itself. That brought trust: if Amazon uses this infrastructure to run its global digital business, so could these startups.

Furthermore, using the free cash flow generated from their core business, they could invest capital into speculative infrastructure which they then offered digital startups as 'infrastructure business angels'. With these offerings, they gave up to $100,000 in free infrastructure to emerging startups. This removed one of the key operating costs for young companies, reducing the cost of failing even further. Anyone who could write some code could now launch a startup and get infrastructure for a year for free (depending on the size of the IT infrastructure required, $100,000 would get some companies even further).

It also ate into the market of Linode, Digital Ocean, OVH, and other hosting companies. And with its free infrastructure offering, it cemented its position as the leader in the digital startup IT infrastructure market. And then they went after SaaS.

When Cloud Infrastructure Became Cloud Services

Now, having built this seemingly infinite pool of digital resources, Amazon realised they could sell a lot more of them if they directly bundled them with applications. If they bundled digital resources with IT application infrastructure components that everyone needed – search, database, content delivery network, virtual desktops, etc. – they could sell more resources while creating convenience for developers. Developers could spin up the components they needed with a click of a button, with no need to be aware of the underlying digital resources required. This concept is closely related to the idea of load-building in the energy sector, as illustrated by General Electric selling fridges (see Figure 4.9).

And as it was, the application infrastructure components already existed as free and open-source software. So AWS began bundling popular open-source products with digital resources, renaming them, and offering them through their public cloud. They churned out services at a remarkable speed, sometimes launching 100 new cloud services (components to create software applications) in a quarter, which they later consolidated again into 'cloud products' (applications themselves, such as email services, business analytics tools, etc.).

All the building blocks that were required for a digital product could now be bought 'as-a-service' with digital resources produced by AWS as a convenient bundle. This not only accelerated the number of digital startups emerging but also increased the scale of AWS as a business significantly. It now represents a majority of the profitability of Amazon's overall business.

It also led to backlash from the open-source community, as AWS often made improvements to the software that would not be given back to the community (as is common practice and often required by many open-source licences).

Figure 4.9 General Electric – demand building by selling electricity-consuming devices to consumers. *Source:* General Electric.

When Microsoft and Google Joined the Party

With the success of Amazon's AWS, it was only a matter of time before other digital behemoths would join the party and they did. Microsoft launched its public cloud called Azure, Google launched Google Cloud, and Alibaba launched Ali Cloud.

But they each had their struggles. Microsoft at that point was not running a large internal IT infrastructure, it was selling licences for customers to run Microsoft's products on their own on-premise infrastructure. They were not in the hosting business, nor a digital-native company, they were a traditional enterprise software vendor. But they had the trust of many enterprise companies and thus their own niche that they could offer their new public cloud to.

So instead of recreating Amazon's market for digital resources, they started on the cloud product and service level and moved their most successful products, Microsoft Office and Microsoft Exchange to the new public cloud, available as-a-service. Through their global reseller network, they began pushing enterprises to shift to this SaaS product, which made much of the IT infrastructure, that enterprises had invested in, obsolete. They even went so far as to stop selling traditional office licences and, instead, bundled them with cloud-based office products (such as email and file storage).

The shift to email- or Microsoft Exchange-as-a-service ('Office 365') did something very important for the later success of Microsoft Azure's cloud infrastructure and services offerings: it punched a hole into the firewall of the enterprise, allowing communication with external Microsoft infrastructure over the internet. This was uncommon, as most enterprises preferred their communications and data to run over their own dedicated or at least fully encrypted networks, with them being the only ones having access. Microsoft managed to convince them that they are as trustworthy and that they can open up their networks to connect with Microsoft's networks. This was like pirates throwing a hook over to another ship.

How important this step was can be seen in the success of Microsoft Teams. Simply put, when remote working emerged, especially during the COVID-19 pandemic, everyone began using video conferencing tools such as Zoom, Webex, Google Meet, Slack, and many others. Corporate users struggled to access any of these services, as the corporate firewall blocked most of these tools. But when Microsoft emerged with Teams, which was free for everyone who already had an Office 365 licence, it was accessible from almost any enterprise network right away – without having to wait for the IT department to make changes to the firewall or 'whitelist' a new service. It propelled Teams to be the most widely used video conference and team chat platform worldwide – within 12 months. Slack especially felt violated and issued a lawsuit for anti-competitive behaviour against Microsoft in Europe.

Unlike Microsoft, Google did have its own, large-scale global IT infrastructure but did not have Amazon's issue of underutilisation. It had purpose-built its infrastructure to be efficient and fully utilised, serving its products, services, and needs. Google recognised early that it was not enough to just collect and store vast amounts of data as a competitive advantage but to also own the matching IT infrastructure to process the data and create value from it. Furthermore, allowing external parties onto its own IT infrastructure posed a risk for Google, a company whose secrecy on how their core products, search and advertising, actually work under the hood is unmatched. So Google went out and either built dedicated Google Cloud facilities or rented colocation space with other providers. At a remarkable speed, it stood up its very own cloud infrastructure to sell digital resources to startups and businesses alike.

However, its niche was less clear. For Microsoft the niche was in the enterprise IT space, it had the trust of companies, and its products were widely used and valued. Amazon had placed its teeth into the digital-native market using discounts and aggressive sales and already owned a significant portion of the market when Google arrived. So Google was left with two options: (i) to ally itself with the open-source community that AWS alienated with its cloud services that didn't give anything back to the community and (ii) to make some of their incredible algorithms available (e.g. for image and voice recognition) on Google Cloud as cloud services to attract a new wave of digital startups focused on machine learning and artificial intelligence.

They did both. And with making Kubernetes available as free and open-source software, they threw a wrench into Microsoft's and Amazon's business model, while creating a massive community of support.

Kubernetes and the Next Layer of Abstraction

With Kubernetes, the IT infrastructure community essentially turned the idea of digital resource primitives into a way of designing and running infrastructure. It enables applications or components of an application to be moved, started, stopped, scaled, and replicated across a heterogeneous IT infrastructure, as long as that infrastructure can produce primitives (compute, memory, storage, and network bandwidth). It created the next layer of abstraction, beyond virtualisation, taking each unit of digital resource and making it allocatable, usable, and programmable by an application.

In practice this means that a Kubernetes-based infrastructure can span across multiple cloud infrastructure providers. It can include infrastructure that sits in an existing data centre or a server room. It can essentially soak up all available digital resources and will dynamically distribute applications (or components of them) based on their needs. Does the component need to be redundant? It will be distributed across three physical servers and will have three instances of it running at all times. Does the application need more resources because of a surge in demand? More digital resources will be allocated or new instances will be started in different data centres while informing the load balancer where to send traffic.

Even further, it enables what Amazon's AWS had done before – to take existing open-source software, such as a database, a CMS, or a search engine, and turn it into a Helm chart (basically an instruction file on how to run this piece of software on the IT infrastructure, what types of resources it needs, how to start and stop it, etc.) and start any service required in a matter of minutes on the Kubernetes infrastructure. It takes cloud infrastructure, its primitive resources, and the services and makes them completely portable, disconnected from the actual cloud infrastructure it is running on.

Now, why does this throw a wrench into the business of the existing cloud infrastructure providers? Because although the use of their digital resources does not create a form of lock-in, the use of their services, which bundle resources and applications do. And this lock-in is what creates a competitive moat for AWS, Microsoft, and others. Once an application is built using the cloud services, to leave a provider, the application would have to be re-engineered. Now if an application is built using Kubernetes as the infrastructure orchestration layer, utilising open-source Helm charts to provide the required services, the only ingredient the Kubernetes infrastructure requires is the digital resource primitives. And those can come from any provider, increasing competition on price and making the majority of the provided cloud services obsolete.

How Traditional Hosting Companies Have Been Left Behind

You are probably wondering what has happened to the industry of existing hosting companies. Weren't they in a sense the first cloud providers when they offered virtual machines? I believe you are right to think that, but it is still a matter of debate.

What is, however, interesting is to observe how the cloud providers have overtaken them so quickly in terms of market share and growth, especially on their own turf: providing IT infrastructure to the actors of the digital economy. To understand this we have to look at the financials behind a hosting and cloud infrastructure business.

To run such a business, five key components are required:

- Data centre facilities (built or rented) and electricity
- Servers (bought or leased)
- Operational staff to run the physical and IT infrastructure (in-house or outsourced)
- Sales and marketing to sell the generated digital resources to customers (directly or via channels)
- Networking and internet access (leased or build)

Financing a new data centre facility or renting space in a third-party facility is relatively cheap compared to purchasing or leasing server equipment (depending on the equipment, it can vary between 5 and 20× the costs). The costs of sales and marketing as well as operational staff are also relatively modest as an operational team of 10 people can often manage an entire hosting business. And the cost of network traffic has come down significantly as well over the years

The real cost driver in a cloud or hosting business are 1) the servers and IT equipment and 2) at the scale of 10+ MW of facility size, the price of electricity. Roughly said, IT equipment is the main capital expenditure (when it's bought) while electricity (at scale) dwarfs the staff costs and is the largest operational expense.

Now what differentiated all of the large cloud providers from the existing hosting companies is that they already had a strong cash-generating business, be it from e-commerce (Amazon), advertising (Google), or licensing (Microsoft). With this cash, they could heavily invest in server equipment (and building data centre facilities) even if they had no customer secured (speculatively); or like in the case of AWS giving significant amounts of resources away to startups for free as a means to acquire them. The willingness to invest up to $20,000–100,000 into customer-acquisition costs (which is what AWS gave away in free credits to them), shows the enormous profit potential of a customer when they ran out of credits.

Why didn't hosting providers simply match this effort? Because buying servers speculatively doesn't work well with the traditional banks that financed those hosting companies. Servers are worthless assets after three to five years, decreasing in value the moment they are bought. Even though as many in the industry know, they often last 5, 8, 10, or even more years, which is when they become very profitable. They were also not large enterprises with financial expertise or stock-listed (albeit OVH, when transforming itself to OVHcloud did list on the stock exchange) to acquire the capital needed to match the cloud providers' growth path.

For Microsoft, it was the same, giving radical discounts of 60–80% over list prices to enterprise customers just to move their infrastructure to Azure, Microsoft's cloud platform, or even going so far as to acquire the existing IT infrastructure of an enterprise (literally paying for 'demand') became common practice.

All of this effort came from the knowledge that once a customer is on a specific cloud infrastructure platform, it becomes difficult to leave. This is likely to change with the widespread adoption of Kubernetes, but for now, the assumption of a cloud provider holds: once you are on it, you are not going to leave. The costs of doing so would be prohibitive. And to make sure it doesn't happen, all of the cloud providers put a lock on the door: inbound traffic (moving data into the cloud) is free, and outbound traffic (moving data off the cloud) costs a fortune.

Table 4.1 Overview egress fees by AWS region.

AWS egregious egress

1 Mbps = TBs per month at 100% utilisation	0.3285
Billed at 95th percentile	0.3458
Estimated average AWS utilisation per month	20%

AWS region	AWS cost at 1 TB/month[a]	Implied Mbps $	Estimated regional cost per Mbps	Markup
US/Canada	$92.07	$6.37	$0.08	**7959%**
Europe	$92.07	$6.37	$0.08	**7959%**
India (Mumbai)	$111.82	$7.73	$1.00	**773%**
Singapore	$122.76	$8.49	$0.50	**1698%**
Korea (Seoul)	$128.90	$8.91	$2.50	**357%**
Japan (Tokyo)	$116.63	$8.07	$0.50	**1613%**
Australia (Sydney)	$116.63	$8.07	$1.00	**807%**
Brazil (Sao Paulo)	$153.45	$10.61	$0.50	**2122%**

Data on current pricing for 1 TB in every region: https://calculator.s3.amazonaws.com/index.
html#key=files/calc-917de608a77af5087bde42a1caaa61ea1c0123e8&v=ver20210710c7. The text in bold shows how much margin AWS is adding to the network capacity they sell vs. what they pay for it.
[a] Taken from the official AWS "Simple Monthly Calculator" as of July 21, 2021.
Source: Cloudflare/https://blog.cloudflare.com/aws-egregious-egress/last accessed December 13, 2022.

Network traffic costs are a massive source of profits for all of the cloud providers as Cloudflare has very well explained (see Table 4.1), also showing the potential costs of leaving a cloud provider:

Hosting companies have been left behind so far due to the lack of access to the capital they can invest in buying servers. This is being partly addressed by OEMs now offering hardware-as-a-service to their clients, realising it is becoming an impediment to the growth paths of their most important customers. As most cloud providers are focused on building their low-cost hardware platforms or are participating in the efforts of OCP to develop low-cost server platforms, OEMs are better off supporting their existing customer base. Likely, the hosting companies will leapfrog cloud providers by jumping on the Kubernetes bandwagon, rather than spending time and effort to recreate the hundreds of cloud services that Azure, AWS, and Google have built to lure customers in.

The challenge, however, will remain, with the cloud providers generating billions of dollars of free cash flow that they can invest into their own infrastructure business (cloud) with yields of 30–40% operating profit (see AWS financial reports). That is a great investment of capital by all means and it is unlikely any of the cloud providers will stop plowing money into the growth of their cloud infrastructure platforms – to acquire more IT demand and to build their infrastructure. After all, why should they share the cake with anyone?

How the New Paradigm of Cloud Infrastructure Is Changing How Data Centres Are Built

Not only has cloud infrastructure reinvented the way a data centre and its servers are divided into digital resources, but it is also taking a stab at the data centre design that has previously been driven by enterprises, hosting companies, and the financial sector. After all, it is a lot more cost-effective to solve redundancy through software (e.g. by running the same application in multiple locations at the same time and load-balancing between them) than through expensive redundancy in the physical infrastructure.

This has already been witnessed in the OCP community designs for servers (see Figure 4.8), where more and more of the responsibility of the data centre facility (backup power and cooling) is migrated into the rack (shifting responsibility from the facility to the IT infrastructure). It can be observed in the latest data centre projects of cloud providers that are leaving out the diesel generator and instead opting for only four hours of backup from batteries, which is likely because they are capable of migrating all of the applications to another facility within that time frame.

What's more is that the business model of the colocation sector is starting to run counter to the motives of future IT and cloud infrastructure:

- Colocation providers sell space, IT, and cloud infrastructure wants to increase density, reducing the space required.
- Colocation providers want to maximise the lifetime of their legacy facilities, IT, and cloud infrastructure providers require new high-density cooling and electrical systems.
- Colocation providers sell electricity with an added margin, cloud, and IT service providers realise that at-scale, purchasing power themselves (as we have seen with large Power Purchase Agreement [PPA] purchasing by Google, Facebook, and Amazon) is a lot cheaper for them.

The paradigm of treating data centres and the servers they house as simple factories for digital resources illustrates the need for fully integrated facilities, which are designed to maximise the total utilisation of the resources that are generated and pay less attention to space utilisation. After all, colocation can be filled with servers that are all turned on and are all counting prime numbers until infinity (doing no useful work).

The profitability of the cloud infrastructure business is driven by the total utilisation of the assets producing digital resources – servers.

There is another important paradigm shift, related to networking, which we will cover in the next section. And another question comes to mind: if most digital startups, most actors of the digital economy either run on their own infrastructure or use cloud infrastructure already, when will enterprise IT fully move to the cloud? That is a question we will try to answer in the next section.

Central Role of Connectivity in the Future of Data Centre Architectures

If you expand the paradigm explained earlier about using networks to unbundle storage and computation from each other, imagine what you could do if all your data centres would all be connected through massive backbone fibre networks. It would drive the paradigm one step further, data could be in one data centre in London, and processing could happen in another data centre in the North of Sweden, Canada, or Norway.

The European Union has called this paradigm a 'cloud continuum', albeit looking at it rather from the perspective of having many smaller data centres superconnected to each other and distributed all over Europe.

Whichever narrative you want to believe in, smaller, distributed data centres all over the place, or medium- and large-scale ones in a few key places, what they all have in common is super-connectivity between each other. Think about it as the electricity grid of the future, a data grid, spanning the globe, all fibre-based, connecting all generators of digital resources to consumers anywhere, computing where digital resources are available, storing data in multiple places to ensure redundancy, etc.

We can see this play out already, again in the realm of the cloud providers, most notably Google and Microsoft, who have been investing heavily in owning their own sea fibre and new land fibre infrastructure. In the case of AWS, it is notable that they can move further and further away from internet exchanges, simply by laying their own fibre connectivity to global internet exchanges. You can see an overview of Microsoft's global network on Figure 4.10.

Over time, however, connecting internet exchanges is likely going to be obsolete, as each cloud provider will become its own internet exchange – after all, if the majority of the digital companies are hosted by three cloud providers, what other connectivity other than to those cloud providers do we need? And likely within each cloud provider's network, all traffic will be free, while traffic outside will be expensive, further creating a competitive moat.

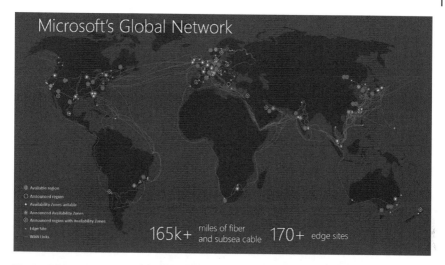

Figure 4.10 Microsoft's connectivity network from the European Commission presentation. *Source:* Microsoft from a presentation by Mark O'Neill for the EU Commission from Wednesday, April 14, 2021.

Problem of an Old IT Philosophy

So when will enterprise IT infrastructure migrate to the cloud? If you ask people in the IT infrastructure of an enterprise, they will likely say 'soon' or 'at some point, we will migrate'. Or enterprises simply hire an external IT consultancy – who is likely a certified partner of one or multiple cloud providers – to do the 'lift-and-shift' for them, likely subsidised or incentivised through the cloud infrastructure providers themselves. For cloud providers, the more demand that can be captured now, the more profitability in both the short and long runs of the cloud infrastructure. This can be compared with the demand-building of electrical utilities (e.g. selling electrical irons) in the 1940s to get people to consume more base-load electricity.

Technically, migrating IT applications should not be that difficult; only it is the chain of dependencies that is often most difficult, e.g. which other applications are dependent on the other, and where are they? To have an overview of the complex IT landscape of an enterprise, the role of the enterprise architect was invented; however, migrating the myriad of off-the-shelf applications, some of which likely ran out of support, or all the custom-made applications can still be daunting. So what happens?

Leave behind and build new! A common philosophy in IT is to simply leave old systems behind, build them again, e.g. using cloud services, tell everyone about the new application, and hope that no one uses the old one anymore. After a few years, the old application can be turned off and the associated infrastructure can be removed. It is IT people who invented tests such as this one: turn a server off (and the applications running on it) and if no one from the business calls within three days to complain, assume nobody is using the applications anymore and remove the server.

There continues to be a vast disconnect between the people that run the IT infrastructure and are responsible to keep it up and running (IT operations or 'ops' people), the people that build the applications (developers and architects), and the people who define the requirements (business analysts, project managers, etc.) and those who use them (business users). This results in the philosophy of just keeping everything running for as long as possible, to the point that the server might actually be dead already, but it is still fed with power and cooling (link to VMWare study on ghost servers).

Yet from a logical standpoint, as long as making digital-only products and services is not the core activity of the business, it's unlikely that in the near future, it will continue to build, manage, and operate its own IT infrastructure. It is much more likely it will rely on the digital resources and services provided by cloud infrastructure providers.

There is a second that underlines this, which is the newfound obsession with data and analytics of most enterprises. As most of them quickly realise, storing data is one thing and processing it is another. In the terminology of the enterprise IT community, if 'data is the new oil', where will you refine it? Many examples point at Google for the mastery of data collection, but Google's true masterpiece is its data storage system and owning the largest data refinery in the world, hundreds of data centres filled with likely millions of servers, ready to compute.

It is highly unlikely that every data-obsessed enterprise will invest the capital needed to build such a data refinery, especially when the business case for storing and processing that data is still written in the clouds (no pun intended). It is much smarter to rent the refining and storage capacity for the time being until it is clear what the business case of this newfound gold mine is. And luckily the cloud providers stand ready with the infrastructure, scalability, flexibility, and services needed to help enterprises launch their new oil-drilling expeditions. There seems to be a strong correlation between this scenario and the California gold rush, with the people selling shovels walking away with the majority of the revenues.

5

Overview of Infrastructure

In order to support the IT hardware, which operates to deliver IT services, a number of physical infrastructure systems are required, which typically include:

1) Power
2) Cooling
3) Monitoring
4) Fire detection and protection/suppression
5) Security

Figure 5.1 is a simplified example of how power and cooling systems support IT equipment. The solid lines show power supplies and the dashed lines cooling. Cooling equipment also needs to be powered and electrical equipment needs to be cooled (in some cases, this is ventilation only without control of temperature or humidity).

There are a number of different design options for each; cost is a driving consideration. Projects are often very focussed on capital cost. However, some cost savings at this stage may be offset by additional costs during operation; for example, low-cost equipment is installed which uses more energy and is less reliable. The total cost of ownership, illustrated in Figure 5.2, takes into account not only capital cost but also operating cost, which is comprised of staff costs, energy, and maintenance costs. A hidden cost is that of failure – if service was interrupted, what would be the cost to the business?

This depends on many factors such as the function of the service, number of users impacted and the duration of outage. For example, an investment bank unable to process transactions for one hour during the trading day could cost the business millions of dollars. However, a one-hour outage on a retail website at an off-peak time may result in some lost sales but many customers will try again later.

Data Centre Essentials: Design, Construction, and Operation of Data Centres for the Non-expert, First Edition. Vincent Fogarty and Sophia Flucker.
© 2023 John Wiley & Sons Ltd. Published 2023 by John Wiley & Sons Ltd.

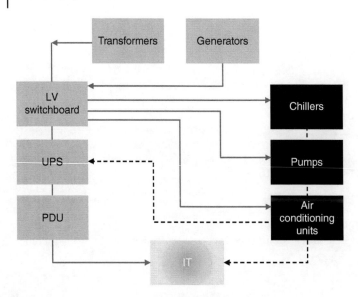

Figure 5.1 Example of electrical power distribution and cooling system using chilled water.

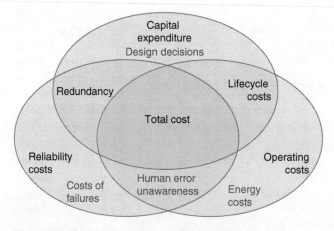

Figure 5.2 Data centre total cost of ownership.

Where the cost of failure is higher, this is usually reflected in an infrastructure design which makes use of redundant components and distribution paths; this spare capacity allows an operation to continue in the event of failure. These additions cost more in terms of capex; the additional plant also has an associated maintenance cost. Often systems with additional available capacity are operated in a way which is less energy efficient, although the two are not necessarily mutually exclusive (see Chapter 9). The reason for investing in higher levels of redundancy is to provide higher reliability and reduce the likelihood of failure. Reliability analysis is a subject in its own right; terminology often used to describe topology is in terms of 'N'. 'N' or 'need' represents the required components to satisfy the design capacity. For example, the design cooling load is 1000 kW; one chiller with a cooling capacity of 1000 kW or two chillers of 500 kW (at the design environmental conditions) is/are installed – there are N chillers. $N + 1$ describes when there is a spare component within a system. In the previous example, $N + 1$ could be two 1000 kW chillers or three 500 kW chillers in the chilled water system. $2N$ is where there is a redundant system. For example, two 1000 kW chillers, each part of an independent chilled water system.

Figures 5.3 and 5.4 help illustrate why $N + 1$ has improved availability compared with N.

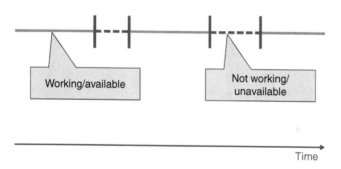

Figure 5.3 Downtime for N system.

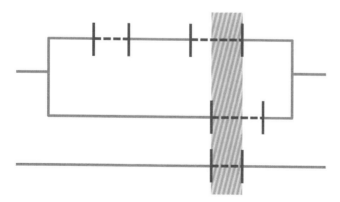

Figure 5.4 Downtime for $N + 1$ system (only when both components are simultaneously unavailable).

Table 5.1 Tier level definitions.

Level	Description
Tier 1	No redundant components
Tier 2	Redundant components
Tier 3	Concurrently maintainable
Tier 4	Fault-tolerant

The data centre industry often describes design reliability in terms of Tier levels. Several tiering definitions exist including EN 50600 availability classes (BSI 2019[1]); most describe the levels as in Table 5.1:

Lower tier levels may reflect a strategy where redundancy is managed at the IT level. So, a loss of power/cooling in one facility does not cause an IT service interruption, as this can continue at another facility.

Tier 3 is typical of many enterprise facility designs and those available for sale/rent. Concurrently maintainable means that any part of the system can be maintained without the requirement for a shutdown, i.e. business services should be able to operate without interruption. The design will include redundant components/systems and the system settings can be adjusted according to a maintenance operating procedure (MOP) to allow maintenance to take place on the required part of the system. Tier 3 designs theoretically should continue operating after a single failure but not necessarily following two simultaneous failures.

Tier 4 designs have redundant systems so that, should any failure result in an entire system being unavailable (e.g. catastrophic chilled water pipework leak), operational continuity can be maintained via the redundant system. These designs also look to physically separate systems, for example, switchboards from two different distribution streams are housed in separate rooms. Therefore, a fire or flood in one switch room would not impact the other. The impact of a high level of redundancy is increased cost and Tier 4 designs are typically only used where there is a high impact in the event of a business interruption, e.g. financial services.

High-reliability designs aim to avoid single points of failure (SPOF) – items which if failed would result in an operational outage.

However, even with redundant components/systems, an unplanned outage can still occur, although theoretically, this should be less frequent than without. Most failures are caused by human error (see Chapter 7).

---

1 BS EN 50600-1:2019 (2019). *Information technology. Data centre facilities and infrastructures. General concepts*. https://doi.org/10.3403/30374947.

In reality, facilities may not be fully aligned to a single tier level throughout; they may have some systems which comply with Tier 3 but others which do not. Or a mixture of Tier 3 and Tier 4 elements, for example, Tier 4 electrical infrastructure which requires continuous power and Tier 3 cooling which can permit a short interruption. The facility is then described as the lowest tier level which applies (a chain is as strong as its weakest link).

It is likely the case that not all IT services within a facility have the same criticality and thus do not all need to be supported by higher tier level infrastructure. The concept of multi-tier hybrid design caters for these differing requirements and allows the reduced sizing and therefore cost of the installed power and cooling systems. However, this is more challenging to manage in terms of predicting demand, managing capacity and operational differences.

Sometimes data centre availability is expressed in terms of 'nines'. Five nines = 99.999% availability. Over one year this would equate to around five minutes downtime. This theoretical figure is not a prediction or guarantee. It might mean one longer outage every few years. Or several short outages which may be more disruptive than one longer one. These figures come from modelling of reliability of components and systems, which can be used to compare the theoretical availability between different topologies. However, it is much harder to model the human element.

Power

IT hardware is typically housed in a rack or cabinet which has internal power strips which the devices plug into. Most IT hardware has two power supplies ('dual-corded') which are fed from different electrical sources to provide redundancy, illustrated in Figure 5.5. In normal operation, the device will draw power from both sources but will continue operating in the event of the failure of one source, by taking its full power requirements from the remaining supply, as shown in Figure 5.6.

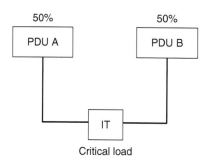

Figure 5.5 Dual power supplies to IT equipment, normal operation.

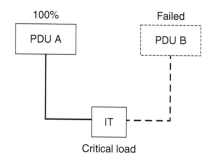

Figure 5.6 Dual power supplies to IT equipment, source B failed.

Power is distributed via cables and/or busbars (solid metal bars which carry current). Cabinet power supplies are fed from a power distribution unit (PDU), an electrical distribution board or a switchboard. Normally, several PDUs are fed from an upstream switchboard. Confusingly, the term PDU is often also used to describe the in-cabinet power strips (particularly by IT people).

Due to the mission-critical nature of data centre applications, in most cases, the electrical supplies to the IT hardware are uninterruptible power supply (UPS) backed, to protect against mains power interruptions. Power blips include not just a loss of power but also other changes to the electrical waveform outside of permitted boundaries, e.g. sags (voltage dip), spikes, and surges.

The acceptable range of operation of IT equipment is defined in the Information Technology Industry (ITI) Council CBEMA (Computer and Business Equipment Manufacturers Association) curve as shown in Figure 5.7, in terms of percentage of nominal voltage and duration. Note the 'No Interruption in Function Region' on the left-hand side of the diagram.

An UPS provides continuous, conditioned power by use of various electronic components connected to a means of back-up power store, which is, in most cases, batteries. Other storage methods exist including flywheels, fuel cells and compressed air.

A typical static UPS is designed to provide uninterrupted power for several minutes when fully loaded. The design may specify 10 minutes of battery autonomy at full load, at N, and at the end of life (battery performance degrades over time), for example.

For short mains power blips, the UPS ensures the connected IT hardware is unaffected. However, if there was a prolonged utility outage to the mains power, the UPS batteries would fully discharge and therefore most facilities also have standby generators which operate to supply power to the electrical distribution in the event of a mains failure.

Rotary UPS systems have a different philosophy where the equipment provides a few seconds' ride through time before the motor/generator, which is continuously spinning, takes up the load.

Many designs provide power to IT equipment via two independent UPS streams at $2N$ (e.g. 1000 kW A and 1000 kW B). Some designs may use distributed redundancy where there are three streams (500 kW A, 5000 kW B and 500 kW C). Each IT rack will still take its power supply from only two streams which could be A & B, B & C, or A & C. This results in capex savings as the total installed capacity is reduced (1500 versus 2000 kW). This is more challenging to manage operationally (including load balancing) due to the increased complexity and failure combinations.

A generator is an engine which produces electricity by combustion of diesel or another fuel. In the event of a mains failure, this interruption in the utility power is detected and the generators are called upon the automatically start; an automatic switching sequence then allows them to start supplying the load, typically in around 30 seconds. Depending on the reliability of the electricity grid, the generator(s) may not be required to operate often. However, as they are vital to

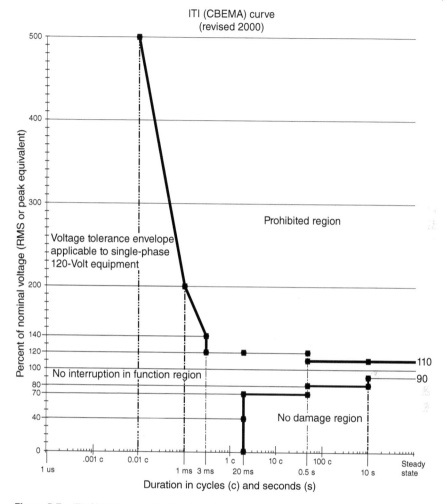

Figure 5.7 ITI/CBEMA curve (**public domain**).

supporting the critical load, frequent testing and maintenance are essential to ensure they remain ready for this event. In some maintenance scenarios, the operational team may manually operate the generators. See Chapter 7.

Many utility companies offer a revenue stream to data centre operators by using their generators to supply power to the electricity grid at times of peak demand, i.e. exporting power. Regular generator operation can be positive for reliability but for many operators, the idea of external connection and remote control of critical equipment is deemed to be too risky. Running the generators has associated emissions which in many cases exceed those from using electricity from the grid.

Cooling fans and sometimes chilled water pumps have a UPS supply (typically via a separate UPS system known as 'mechanical UPS' to indicate that it feeds mechanical equipment). This is used to help support continuous cooling during a power failure. However, any equipment which is not UPS-backed, such as chillers, will see a power interruption during a mains failure. Once the generators start supplying the load, they should restart automatically. The temperatures of air/water supplied to IT equipment will increase as a result of this cooling interruption, the extent of which depends on several factors including load density, thermal inertia, and segregation of hot and cold air streams. This is described as a 'thermal runaway'. By design, this temperature increase should not exceed the defined high-temperature limits. Some chilled water designs incorporate buffer vessels as a thermal store which discharges chilled water whilst the chillers are not operational. This increases the mass of chilled water contained within the chilled water system and is designed to increase the inertia.

Designs may incorporate electrical changeover devices to allow an alternate source to be used to supply the connected load in case of failure of the primary source, as described in Table 5.2:

Note that the changeover device itself can be a point of failure.

When mains power becomes available again, there is a further transfer from generator power back to mains. This may be automatic or manual at a time chosen by the operator. The operator may choose to wait until grid stability has been regained (sometimes the first power interruption may be followed by others) and undertake this switchover back to mains at a lower risk time of day, depending on the perceived risk to business continuity if there is an unplanned failure during this activity. The transfer back to mains may be designed as a break transfer (open transition) where the UPS batteries support critical loads and non-UPS loads lose power again; or a no-break transfer (closed transition) where the generators are briefly paralleled with the mains and no further interruption is seen by the load. For open transition, there will be a further interruption to cooling (where not UPS-backed), so this transfer is best avoided until the temperatures have recovered following thermal runaway.

Table 5.2 Types of changeover device.

Changeover device	Nature of changeover	Typical application
STS (static transfer switch)	No-break	Single-corded IT equipment
ATS (automatic transfer switch)	Short-break	Cooling equipment
MTS (manual transfer switch)	Longer break	

Cooling

The electrical energy supplied to the IT hardware is converted into heat and therefore a cooling system is required to remove this heat and keep the IT hardware at an appropriate temperature. In addition, the cooling design may make an allowance for other heat gains in the space (thermal gains through the ceiling, walls, lighting, etc.); sometimes this is considered negligible. Heat densities in data centres vary however, the trend is increasing.[2] These are much higher than in a typical office, for example, one server might have a load of 1 kW; with 10 in one rack, the density is 10 kW in $0.6 \times 1.2\,m^2$ (typical rack footprint). A portable electric heater could be 2 kW by way of comparison. A row of 15 racks would therefore be 150 kW. By comparison, a human and their computer and monitors could have a heat load of around 200 W (around 1 kW in an equivalent area). Some racks may be described as 'high density'; however, there is no globally accepted definition for this. Supercomputers or high-performance computing applications would typically fall into this category.

Most air-cooled hardware takes air in the front and rejects warmer air at the back. Exceptions may include networking equipment such as switches. Rows are therefore organised in a hot aisle/cold aisle configuration (front to front and back to back) to help separate hot and cold air streams.

Typically, air-conditioning units, also known as computer room air handling (CRAH) units or computer room air conditioning (CRAC) units, cool hot air from the IT equipment and supply cooled air. There are many variations on cooling design, which comes in a range of shapes, sizes, and locations inside/outside the space: air may be delivered by air handling units (AHUs), indirect evaporative coolers (IECs), in-row coolers (IRCs), rear door coolers (RDCs) . . . The term close-control unit (CCU) used to be more prevalent but current best practice accepts a wider range of conditions, so close-control is not required. Cold air may be supplied from a raised floor via grilles, as illustrated in Figure 5.8.

IT hardware does not require the same environmental conditions as humans in an office. The main IT hardware manufacturers through ASHRAE's Technical Committee 9.9 for mission-critical facilities, data centers, technology spaces, and electronic equipment (TC9.9) work together to define common thermal guidelines for temperature and humidity. These guidelines state the ranges of temperatures and humidities which can be provided at the inlet of the IT equipment so that it can operate without affecting its warranty. Exceeding the maximum temperature limits can void the warranty and may result in premature failure of hardware. The current

---

2 ASHRAE Technical Committee 9.9 (2018). *IT Equipment Power Trends*, Third Edition. ASHRAE.

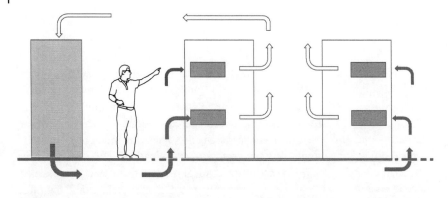

Figure 5.8 Raised floor cooling with air flows in section (hot aisle-cold aisle).

Table 5.3 ASHRAE 2021 air cooling environmental guidelines recommended and allowable A1 ranges.

Recommended for low pollutant levels (<300 Å/month Cu, <200 Å/month Ag)	
Temperature	18–27 °C (64–81 °F)
Humidity	5.5 °C (42 °F) dew point/70%RH & 15 °C (59 °F) DP
Recommended for high pollutant levels (>300 Å/month Cu, >200 Å/month Ag)	
Temperature	18–27 °C (64–81 °F)
Humidity	5.5 °C (42 °F) dew point/50%RH & 15 °C (59 °F) DP
Allowable A1	
Temperature	15–32.2 °C (59–90 °F)
Humidity	20–80% RH/17 °C (63 °F) DP

guidelines for air cooling were published by ASHRAE in 2021[3] and define recommended and allowable ranges of operation. These are summarised in Table 5.3.

The trend for the environmental conditions has been widening, however in 2021 a further class was added for high-density air cooled IT equipment H1 with a lower upper limit (maximum 25 °C allowable and 22 °C recommended). Note that altitude also derates the maximum temperatures.

These environmental envelopes may be represented on a psychrometric chart,[4] as shown in Figure 5.9. A detailed technical explanation of psychrometrics is beyond the scope of this publication.

3 ASHRAE (2021). *Thermal Guidelines for Data Processing Environments*, Fifth Edition. ASHRAE. https://www.ashrae.org/technical-resources/bookstore/datacom-series.

4 Gatley, D.P. (2013). *Understanding Psychrometrics*, Third Edition. ASHRAE.

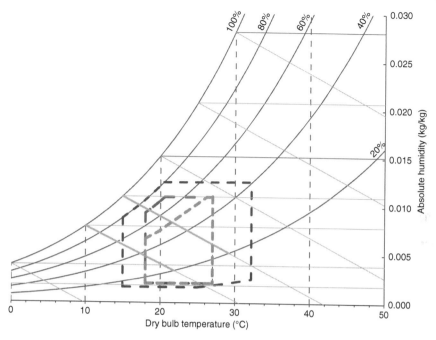

Figure 5.9   Psychrometric chart with ASHRAE 2021 recommended and allowable A1 ranges.

The psychrometric chart is particularly useful for discussions around humidity requirements. Humidity is often misunderstood in part due to how it is described (absolute humidity/dew point temperature/relative humidity). For a given absolute humidity or dew point temperature, relative humidity varies with dry bulb temperature due to the physical properties of air (hotter air can hold more moisture).

The environmental envelopes are revised from time to time; the reason for their broadening in the last few years has been to allow energy savings in cooling systems. It requires more energy to cool to a lower temperature. However, there has been a perception that colder IT hardware is more reliable. There is a correlation between the operating temperature and hardware reliability. However, TC 9.9 published an analysis which indicated that in most climates, any increase in failure rates with elevated operating temperatures was not statistically significant.[5]

---

5 ASHRAE Technical Committee 9.9 (2011). *2011 Thermal Guidelines for Data Processing Environments – Expanded Data Center Classes and Usage Guidance* Whitepaper prepared by ASHRAE Technical Committee (TC) 9.9 Mission Critical Facilities, Technology Spaces, and Electronic Equipment, ASHRAE.

The reason for having a recommended range and allowable ranges is that most IT equipment operates at a higher fan speed at higher inlet temperatures. Acoustic tests for IT equipment are undertaken at 25 °C; the equipment is therefore operating at a minimum (quiet) fan speed at this temperature. At higher temperatures, the on-board fans may increase their speed but note that this may be at much higher temperatures (above 35 °C), depending on the hardware. If the IT equipment was operating all the time in the allowable range, there is a risk that:

1) The noise levels in the room could require operators to wear hearing protection in order to enter the space.
2) The savings from operating the cooling systems warmer may be offset by the increase in power draw by the IT equipment fans. However, if this scenario occurs for only a limited time period, higher temperature operation is still favourable overall.

Another common misconception is the location of these temperature and humidity requirements. Often the perception is that all areas of the data hall must comply with these environmental conditions. A human in a data hall operating at the upper end of the recommended range is likely to find the environment much warmer than a comfortable office space, particularly in the hot aisle. However, the critical temperature is what is supplied at the inlet of the IT equipment. Servers and other IT hardware typically have their own fans and internal heat sinks which manage the internal cooling of the equipment. The hardware rejects heat at its outlet, which means that this air is hotter than at the inlet. How much hotter depends on the hardware and workload; this varies between a few degrees to around 25 °C hotter. This means the hot aisle can be very hot but this is by design and does not present an issue for the IT hardware. There are other considerations for high hot aisle temperatures which require managing, such as electrical cable ratings and the health and safety of operators. Many legacy installations control the return temperature of their room cooling units to 22 °C, which is typical for cooling an office. The resulting critical IT equipment inlet temperature is often very different to this controlled temperature, as illustrated in Figure 5.10. A better strategy is to control the supply air temperature from the cooling units. Where good air management is in place, this should be close to the critical IT equipment inlet temperature.

Where air is not managed, a large amount of recirculated and bypass air is present.[6]

---

6 Hwaiyu Geng, P.E. (2021). Chapter 34 – Data centre air management. *Data Center Handbook: Plan, Design, Build, and Operations of a Smart Data Center*, Second Edition. R. Tozer and S. Flucker (Eds.). Wiley.

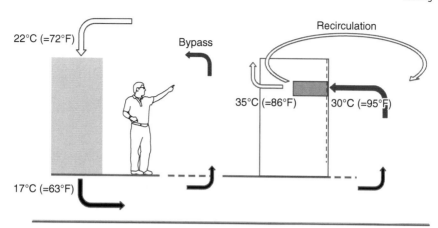

Figure 5.10   Raised floor cooling with air flows and temperatures in section.

Recirculation is where warmer air expelled from IT equipment re-enters IT equipment (itself or a different device). This leads to elevated inlet temperatures and hot spots – where this is excessive it is a risk to hardware reliability. Often this is offset by lowering operating temperatures, which means increased energy consumption.

Bypass air is supplied by the cooling units but does not pass through the IT equipment and therefore does not provide useful cooling – the cold air is going to the wrong place. Cooling unit fans are moving excess air; another wasteful use of energy.

To better manage the air, physical segregation of hot and cold air streams is required. Typically, this is done with some type of containment system, illustrated in Figures 5.11–5.13.

As temperatures in data centres increase (operating set points and IT equipment delta T), the hot aisle temperature is becoming hotter. The chimney rack solution limits workers' exposure to high temperatures and therefore has advantages in terms of managing worker comfort.

It is important to note that the purchase and installation of a containment system does not guarantee good air management and segregation of hot and cold air streams. If it is full of gaps, e.g. blanking panels missing in racks, missing side panels, and lack of cable brushes or gaps in the raised floor, this allows air mixing. The ongoing management of the space must be maintained to keep bypass and recirculation under control.

Cooling is not just a matter of controlling temperature. Air volume is the other essential element. The design and operation of the cooling system should ensure that at least as much air is supplied to the IT equipment, compared with the volume it requires. In most cases, air volume is oversupplied, resulting in wasted cooling equipment fan energy. Managing bypass air is therefore an enabler for fan energy savings.

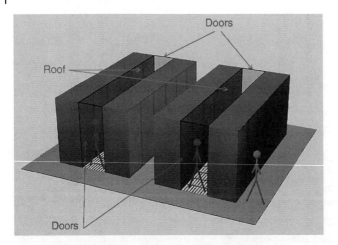

Figure 5.11   Cold aisle containment with raised floor.

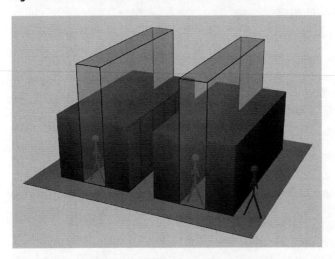

Figure 5.12   Hot aisle containment without raised floor.

Computational fluid dynamics (CFD) modelling may be used as a tool to predict temperatures, air volumes, pressures, etc., within a data hall. This can be particularly useful at the design stage to compare alternative solutions, especially where these are new/unproven. The results can be presented in a graphical format which is easy to interpret for a non-technical audience. For existing facilities, it is often more effective to analyse measured data rather than build a complex model to represent a dynamic environment accurately through survey; the latter can be a resource-intensive process and needs updating to reflect changes to the installation such as hardware installs / decomms.

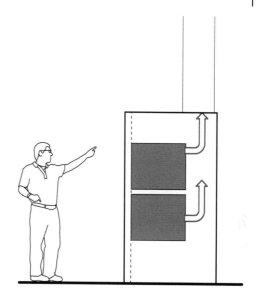

In order to supply air for IT equipment at the required environmental conditions, in most cases, some kind of air conditioning system is employed. The equipment used is in many cases similar to office air conditioning, for example, via a chilled water system or direct expansion (DX) split refrigeration system. There is a heat exchange between the air and the cooling coil in the CRAC/CRAH unit. This is then rejected directly via the condenser for a DX system or there is a further heat exchange between chilled water and refrigerant in the case of a chilled water system (in this case the evaporator is in the chiller itself).

The refrigeration cycle, as illustrated in Figure 5.14, comprises a compressor, expansion valve, evaporator, and condenser; in driving the refrigerant around this process, where it is converted from gas to liquid and back again, heat is removed from the evaporator and rejected at the condenser.

Refrigeration energy savings can be realised with higher evaporator temperatures and lower condensing temperatures. Evaporator temperatures are dependent on the air temperature set points (and water temperature setpoints where applicable); condensing temperatures are governed by the ambient temperature. Reducing recirculation within the data hall enables refrigeration energy savings as this helps ensure that temperatures are managed, thus allowing setpoints to be increased.

As with a domestic refrigerator, electrical energy is required to drive the cycle (via the compressor) and bring down the temperature. In this example, heat is rejected at the back of the fridge into the room but for higher-capacity cooling systems, the heat must be rejected to atmosphere.

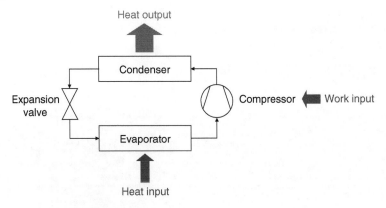

Figure 5.14 Refrigeration cycle.

Factors affecting the choice of the system include the local climate and available space for the cooling plant. Many refrigerants are HFCs (hydrofluorocarbons) which are subject to the Kigali Amendment to the Montreal Protocol[7]; an international agreement to gradually reduce the consumption and production of HFCs due to their global warming potential. Alternative refrigerants often have higher flammability and thus require stricter processes when handling.

Many modern facilities now employ some type of free cooling (also referred to as economiser cooling). This means that the cooling is achieved without running compressors, making use of ambient conditions for cooling/heat rejection instead. The cooling is not entirely free, only compressor-free; energy is still required for the fans/pumps where applicable.

Often it is assumed that free cooling means that outdoor air is supplied into the data hall for cooling. This is one type of free cooling (direct air) but there are a variety of other types, summarised in Table 5.4.

Many of these designs make use of adiabatic evaporative cooling; cooling through evaporating water is less energy intensive than compressor cooling. The amount of water consumed on site can be high, though there may still be a net water saving when considering the water consumed in the process of generating electricity (via a steam turbine, for example). For any system using water, control of water quality is an important consideration, both in terms of hygiene and scale formation, which can impact reliability and capacity. The designer needs to

Table 5.4 Summary of free cooling system types.

Type of free cooling	Pros	Cons
Direct air	Simple (less components), lower TCO particularly if zero refrigeration	Space requirements Management of pollution (potentially high filter consumption)
Indirect air	Allows low-energy cooling in many locations including hot dry climates	Space requirements. May require high water consumption
Water, e.g. CRAH units with dry/adiabatic coolers or cooling towers for heat rejection	More flexible use of space	Less efficient than air-free cooling
Pumped refrigerant	More flexible use of space, decentralised system	Less efficient than air and water-free cooling. Currently proprietary products from one manufacturer

consider how to avoid stagnation in dead legs, for example and the operations team need to implement a water quality management regime.

There is a perception that very cold ambient conditions are required to benefit from free cooling. However, the ASHRAE recommended and allowable ranges permit relatively warm temperatures. Some data centre cooling designs are zero refrigeration, which allows a capital cost saving (including in sizing the electrical infrastructure) and maintenance cost saving as well as energy saving.

The climate is an important factor in selecting and sizing cooling equipment (and thus the electrical equipment which supports it). For a given location, local weather files can be obtained by the designer, which record historical climate data including temperature extremes. Design conditions can be defined based on this information (typically part of the Employer's Requirements). However, climate change means that increasingly, historic weather data is not a good predictor for the future. Tools are available to assist with the modelling of likely future scenarios.[8] Clearly, it's important that cooling systems can continue to operate through a heatwave; the challenge is to avoid wasteful oversizing of equipment. Some considerations:

---

8 CIBSE (2009). *Technical Memorandum 48: Use of climate change scenarios for building simulation.* CIBSE. ISBN: 9781906846015.

- Will the facility ever operate at 100% load? If not the cooling system has an in-built capacity which may be able to absorb these additional demands. Higher ambient conditions may effectively downrate the installed equipment.
- What is the impact if the ambient conditions exceed the design? The indoor temperatures may start to exceed their target limits. Is this a problem? It depends on how much and for how long. If the impact is occasional excursions from the recommended to the A1 allowable range then there is no impact on the equipment warranty. If there is an SLA penalty payable to tenants, then perhaps this is what needs to change, rather than the design.

Further information on best practices for energy-efficient design and operation and low embodied impacts may be found in Chapter 9.

Some IT hardware is liquid-cooled; there are limits to the heat density which can be cooled by air and liquids have higher thermal transfer properties. Liquid cooling currently has a small market share and is used primarily for high performance computing applications.

However, if the forecast IT hardware densities are realised, the dominance of air cooling will be replaced by a liquid cooling era[9] and ASHRAE recommends that any future data centre includes the capability to add liquid cooling into their design. See Chapter 11.

BMS

The Building Monitoring System or Building Management System gives the operator visibility of how the plant and systems are working and can provide alarms in case of faults. This information can often be interrogated via a software interface in a control room. Users may also have remote access.

Mission-critical facilities often use the philosophy of 'global monitoring local control'. With control comes risk – something which can control could potentially cause an unintentional shutdown. Control systems should be designed to 'fail-safe'; for example, if control signal is lost, the fallback position is to continue operating.

Fire

Fire design is driven by local code and insurance requirements.

The specification may require different fire ratings for different zones, for example, a room with a two-hour fire rating is capable of containing a fire for two hours.

---

9 ASHRAE Technical Committee 9.9 (2021). *Emergence and Expansion of Liquid Cooling in Mainstream Data Centers*. ASHRAE.

Spaces need to be reasonably well-sealed (i.e. not too leaky) in order for the fire protection system to work effectively, i.e. not too much air ingress/egress. A room-integrity test is used to demonstrate this and the insurer may require this to be repeated on a regular basis (e.g. annually) after handover.

There are some differences in the application of fire detection and protection in data centres compared to other building types.

The design philosophy usually differs from that of fire systems for people buildings which focus on life safety. Data centre fire design typically avoids interlocks to shut down other systems (because of the impact if the shutdown is triggered erroneously) and allows running to destruction. Fire detection may result in a shutdown of the fresh air supply to a space to avoid adding oxygen to the fire, but not the cooling system (note than in a direct air-side free cooling system the two are combined).

Gas suppression, water mist and oxygen reduction fire systems are often employed rather than sprinklers to limit damage to IT equipment. Due to the mission-critical nature of the environment and the high value of IT equipment, fire suppressant discharges which cause damage/downtime can be costly.

Fire detection may make use of smoke and heat detectors as well as high-sensitivity smoke detection (HSSD), which is a very sensitive 'sniffer' that alerts operators to the very early stages of a fire in development, allowing them to investigate before any of the detectors are triggered.

Security

The level of security will depend on the client's requirements with specific standards applicable to military or government security facilities, for example; these may include background security checks and bag searches. Many facilities do not display company logos outside the building to keep a low profile. Government ID such as a passport or driving licence is often a requirement of entry, along with an approved attendance ticket. Unescorted access may be granted on a time and area-limited basis to specific visitors. Biometric readers are often used in addition to badge readers to control access.

6

Building a Data Centre

Stakeholders, Design, Construction, and Commissioning

Stakeholders and Project Stages

Data centres are a complex ecosystem with many stakeholders from different disciplines interacting at different phases of the facility lifetime. How successfully these stakeholders communicate and share information is essential to the performance of the facility.

The data centre's lifecycle comprises different stages. In the United Kingdom, the RIBA *Plan of Work*[1] indicated in Figure 6.1 is commonly used to define building project stages.

There are several key interface points between project stages and stakeholders, where information and knowledge are transferred[2]:

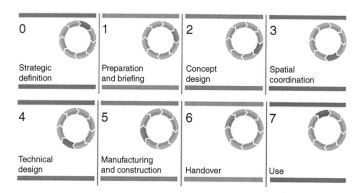

Figure 6.1 Project stages extracted from RIBA's plan of work.

---

1 (2020). *RIBA Plan of Work 2020 Overview*. RIBA. www.ribaplanofwork.com.

2 Whitehead, B., Tozer, R., Cameron, D. and Flucker, S. (2021). Managing data center risk. *Data Center Handbook*. H. Geng (Ed.). https://doi.org/10.1002/9781119597537.ch7.

Data Centre Essentials: Design, Construction, and Operation of Data Centres for the Non-expert, First Edition. Vincent Fogarty and Sophia Flucker.
© 2023 John Wiley & Sons Ltd. Published 2023 by John Wiley & Sons Ltd.

Item	Interface	Purpose
Design brief	Client/designer	Outline requirements/objectives
Specifications	Designer/contractor	Define requirements for installation and operation
Handover record information including Operating and maintenance (O&M) manuals, record drawings, commissioning records, training	Contractor/operator	Reference material for what is installed and how to operate it
Post-handover review	Operator/project team	Feedback on how the facility operates in reality and how this might be improved from an operability perspective. Lessons learned for future designs.

The O&Ms often do not contain information about the design philosophy as they do not contain the original design brief and specifications. A useful document to be updated through the project and handed over to operations is the Basis of Design, which summarises the key elements of the installation and how they are designed to work.

Pre-project

Before there is a project, a need will be identified. This might be customer-driven, for example, demand for IT hosting in a particular location. For an enterprise, the requirement may be due to an existing facility which is approaching the end of life. Data centre operators may pay a provider to build a facility on their behalf, particularly in a new location (there may be programme, risk, and cost benefits).

The data centre owner may need to seek investors to finance the project. See Chapter 8.

Location is typically defined fairly early in the project. See Chapter 3.

Key stakeholders:

- **End user(s)**
- **Company board**

- **Head of sales**
- **Economist**
- **Investors**
- **Local government**
- **Utility companies**

Pre-design

The client will develop a design brief which outlines what the project needs to achieve. These may be known as **Employer Requirements** (ERs) or **Owner's Project Requirements** (OPRs). It is important that these reflect the end user requirements. Including more detail at this stage helps focus the design on how the requirements will be met. Key items to define at this stage are:

- IT load in kW
- Location
- Data hall/building size
- Redundancy and resilience, e.g. concurrently maintainable
- Performance expectations, e.g. target PUE
- Programme
- Budget

Determining priorities assists with forming the procurement strategy and helps inform decision-making as the project progresses.

The client may also define details of the power and cooling systems to be used. Some clients have global standards and template designs which define in detail their requirements. However, it may also be the right time to review previous designs and consider whether other solutions are worth exploring; the end user requirements, market demands and technological solutions are not static.

One of the challenges in designing power and cooling systems is that their lifecycle is many times longer than that of IT equipment. IT hardware is typically refreshed every three years and the new generation of hardware may have different requirements in terms of load density, for example. Electrical and mechanical equipment may have a lifespan of five to seven times longer than IT hardware; it is difficult to predict the IT requirements for the next generation of hardware let alone that in 15–20 years' time. Clients and designers may try and mitigate this uncertainty by keeping up to date with emerging trends and incorporating flexibility and modular aspects into their designs.

Classifications, Standards, and Certifications

There are various certifications and standards which apply to data centres. One of the first was the Uptime Institute's tier levels[3] which define increasing levels of infrastructure redundancy using tier levels I–IV. There are other similar definitions such as TIA's 942 standard.[4] Both offer schemes to independently certify that a facility complies with a particular infrastructure level. EN50600 is a European standard which describes availability classes in a similar way.[5] A description of tier levels can be found in Chapter 5.

Once an initial budget is approved and funds secured, resources will be mobilised including:

- **Project management**
- **Quantity surveyor/Cost consultant/Contract administrator**
- **Procurement**
- **Legal specialists (See Chapter 10)**

There may be a requirement to apply for planning permission with the local authority. This process can include engagement with neighbours and restrictions may be imposed such as building maximum height and sound levels.

Every project is different and it is important to ensure that the are no scope gaps which leave important tasks unaccounted for. The development of a roles-and-responsibilities matrix early in the project can be useful. Examples are provided in BSRIA's *Design Framework for Building Services.*[6]

Typically, there will be a tender process in order to appoint the designer/design and build contractor.

Design

The data centre industry includes several mature operators with defined data centre building requirements. These clients have in-house engineers across multiple disciplines and so have the capability to specify the requirements fully, including the expected facility performance (e.g. PUE). It is often the case that the

---

3 uptimeinstitute.com (n.d.). *Data Center Tier Certification.* [online] Available at: https://uptimeinstitute.com/tier-certification.
4 ANSI/TIA-942 (2017). *Telecommunications infrastructure standard for data centers* (12 July).
5 BS EN 50600-1 (2019). *Data centre facilities and infrastructures.*
6 Churcher, D, Ronceray, M and Sands, J. (2018). A Design Framework for Building Services BG6/2018, Fifth Edition. BSRIA.

employer's requirements have defined engineering solutions to the concept or schematic stage, allowing the local designer or contractor to pick up the detailed design to carry out the employer's engineering intent.

The designer will produce a design based on the requirements, which outlines their proposal for how to deliver the brief. Several different options may be presented for review by the client. This typically includes sufficient detail to size equipment and plant rooms. The client is typically expected to sign off the design at key stages.

Depending on the contractual set-up, the designer may be employed directly by the client or in the case of a design and build contract, the designer forms part of the contractor's team (the role may still be outsourced to an external design consultant).

The detailed design provides enough detail for construction. Many projects make use of BIM[7] (Building Information Modelling), a three-dimensional (3D) modelling process where project stakeholders share information about the installation as part of the design, planning, and management processes in the construction lifecycle and after handover. Each asset will exist in the model with its key information attached. 3D models help different stakeholders coordinate services as working on a common model allows physical clashes to be quickly identified. These can then be resolved prior to installation rather than finding out during the build phase when rework has a potential cost, time, and waste impact. Specifications will be produced to define the requirements of the various plant and equipment and allow them to be procured.

After the specifications have been finalised, requests may arise for changes. These should follow an auditable change management process so that deviations from the original requirements are properly tracked. Changes can have programme and budget impacts.

Key stakeholders:

- **Architect**
- **Civil engineer**
- **Structural engineer**
- **Design consultant – Mechanical and electrical engineers**
- **Building control**
- **Main contractor/General contractor**
- **IT solution/network architects**

Building projects are traditionally architect-led. However, in data centres, the building services (power and cooling systems, etc.) are a more critical and costly

---

7 ISO 19650 *Building Information Modelling (BIM)*. https://www.bsigroup.com/en-GB/iso-19650-BIM/.

part of the project compared with the architecture and thus the engineering consultants have a more significant role. The architect still has an important role in designing the flow of people and equipment through the space and sizing the rooms. In addition to the data halls where the IT equipment is housed, most facilities also incorporate the following:

- Network room/Point of Presence (PoP) room/Meet Me Room (MMR) – where the network enters/leaves the building providing connectivity between the outside world and the data hall(s).
- Plant rooms to house electrical, mechanical, and other building services equipment. May include secure outside spaces, e.g. roof, external compound, and containerised plant rooms.
- Build room/Debox room – where IT hardware is removed from packaging and prepared for installation/where it is stored when removed from the data hall.
- Loading bay – where IT hardware and other equipment enter/leave the building. Access for long vehicles needs to be considered.
- Goods lift/passenger lift.
- Network Operations Centre (NOC)/operations team offices/client offices.
- Breakout room and welfare facilities (toilets, showers, and changing rooms).
- Reception/security office.

Purpose-built data centres typically have an increased slab-to-slab height compared with a standard commercial building. This is to cater for the higher load density; space is needed for cabling and air flow. Most legacy facilities use a raised floor for services distribution, with cooled air delivered via floor grilles. Some make use of a suspended ceiling for hot air return/services distribution (with or without raised floor). In recent years, the trend has been to move away from raised floors – this has various benefits in addition to cost savings. However, this means that all services must be distributed overhead.

Data centres also typically have increased structural requirements compared with standard commercial buildings. Each IT rack loaded with servers, etc., can weigh around 1 ton. Sometimes spreader plates may be added under a raised floor to support particularly heavy pieces of equipment.

Liquid cooling of IT equipment may have floor design impacts:

- For water distribution to the rack, underfloor distribution is usually preferred to better contain leaks.
- Liquids used for this application are denser than air and so are heavier. When using submersion cooling in particular (where there is a relatively concentrated volume of liquid), it is important to understand the structural requirements.

These specific requirements can be challenging to overcome when converting other building types into a data centre facility.

Often the design will be peer-reviewed by an independent consultant, the objective being to ensure the design fulfils the brief. The exercise involves checking the design against the employer's requirements and highlighting any gaps or improvements, which may be more easily rectified at an early stage of the project.

A high-performing design, which is easy to build, commission, and operate with a low total cost of ownership is usually a simple one. This is particularly important given the skill shortage in the industry (see Chapter 7).

Bricks and Mortar versus Modular Build

Data centre construction may be via traditional onsite construction ('stick build') or the provision of offsite constructed modules. One advantage of modular construction offsite is that quality control is easier to achieve in a factory environment. Another is that it allows concurrent construction activities to be progressed which therefore results in programme savings of between 30 and 50%[8] which may result in lower costs. For instance, manufacturing can occur in parallel with foundation work, unlike the linear timeline of a traditional project. A lean offsite manufacturing process can be faster and less wasteful than the equivalent building process onsite. This efficiency is due to the enclosed and controlled factory environment, the ability to coordinate and repeat activities, and increasing levels of automation. The number of shifts also impacts capacity and output times; typically, two 8-hour shifts are used, although if the appropriate labour is found, three shifts could, in theory, be possible.

Modular construction is favourable where the structure has a degree of repeatability, a unit size that suits land transport and a value density where the savings of shifting activities to the factory outweigh logistics costs. Data centres often have repeatable modules; the data halls, cooling and power modules, and fire suppression all have repeatable features. Units may be shipped with a high level of completion with just connections to be made on arrival to the site once moved into position on skids. There are also fewer costly weather-related delays. Production-controlled repeatable processes may have lower costs associated with quality control. Any manufactured module needs to be designed for the manufacturing process. Thus, it is desirable to limit the number of variations.

By manufacturing modules offsite, the factory affords greater control over the working environment. Production line techniques often provide a safer working environment, eliminating work at height, reducing noise, exposure to ultraviolet

---

8 Lawson, M., Ogden, R. and Goodier, C. (2014). Design in Modular Construction. Boca Ratón: CRC Press.

rays, working in confined spaces, and congested work with trades overlapping. Higher quality can be secured due to a controlled environment and thorough testing of modules before delivery and assembly.

Also, because much of the construction and assembly work is carried out offsite, building sites are safer, quieter, cleaner, and generally less disruptive. These advantages are significant when data centre projects are situated next to offices and residential areas. Logistical management of deliveries and just-in-time sequential delivery of modules is key. If a module arrives damaged or out of sequence, the construction and cost impacts may be significant.

Procurement

Perhaps the first construction procurement decision is whether to have an employer's design with a separate construction contract or contracts. It is often the case that the data centre procurement is based on design and build. This puts the responsibility for both elements with one party and may benefit from programme and cost efficiencies.

There are many forms of contract used globally for the construction of data centres. FIDIC,[9] AIA,[10] and NEC[11] all have the potential to deliver the right outcomes for data centre clients and contractors. Many in the data centre industry may view some of these more traditional forms of contracts as having negatives, in that they can create a non-collaborative and combative environment and limit client choices. There are options that are potentially more collaborative such as PPC2000,[12] the first Framework Alliance Contract,[13] FAC-1, and the first Term Alliance contract. FAC-1 is a particularly versatile standard form framework alliance contract which enables a client and their team to obtain better results from a framework, helps integrate a team into an alliance and also helps obtain improved value through building information modelling.

9 fidic.org (n.d.). *FIDIC | Why use FIDIC contracts?* | International Federation of Consulting Engineers. [online] Available at: https://fidic.org/node/7089.

10 Aia.org (2017). *Welcome to AIA contract documents.* [online] Available at: https://constructiondocuments.aia.org/Products/PayGoCategoryProducts.aspx [Accessed 1 Nov. 2019].

11 Neccontract.com (2013). *Contracts, project management & procurement - NEC contracts.* [online] Available at: https://www.neccontract.com/.

12 Project Partnering Contracts and Alliance Forms from the ACA (n.d.). *Project partnering contracts and alliance forms from the ACA.* [online] Available at: https://ppc2000.co.uk/ [Accessed 12 Nov. 2022].

13 King's College London Centre of Construction Law *FAC-1 – Alliance forms.* [online] Available at: https://allianceforms.co.uk/fac-1/.

Whatever the contract form, a successful data centre project requires proper contract documentation. Having contractual protections put in place before the commencement of the project will limit future risk and, hopefully, provide for the smooth construction of the facility.

Developing a greenfield data centre comes with its own risks and challenges. The developer must decide on the right approach for such a development: delivering the project with multiple package contractors and a potential designer/engineer or choosing a turnkey approach whereby an EPC contractor delivers the whole project and agrees to engineer, procure, and construct the data centre. While the first option may provide a more cost-efficient solution, a turnkey EPC contractor undertakes a highly complex project's full completion, turnkey, and interface risk. One of the most obvious benefits of entering into a design and build contract is having one point of contact and responsibility for the project, thereby avoiding having to manage the various stakeholders involved in construction and project delivery.

Contract terms may be common with other types of construction. However, nearly all data centre contracts come with some of the following in various shapes or sizes:

A non-disclosure agreement (**NDA**) is typical because there is a need to share information even in the earliest stage of a data centre project and it is important that it is not assumed that conversations with advisors are confidential. For the purpose of evaluating the potential for a future relationship, NDAs are commonly executed when two parties are considering a relationship and/or collaboration together and need to understand the other's processes, methods, or technology.

A **force majeure** clause regulates how a party's contractual obligations may be altered due to an event beyond their control. Recent supply chain issues, prevalent in many of today's construction projects, may impact the sourcing of necessary equipment. There is also the issue of climate impacts and what should and cannot be anticipated. Depending on how this provision is drafted, a force majeure event may include various scenarios that may result in increased costs, delays, or even a possible termination event.

A **scheduling and liquidated damages provision** if the project is delayed may reduce the financial impacts associated with a late completion. The damages may be extensive; there is a whole range of damages that may follow from a late completion including the fit out of racks and data mitigation, and loss of operational revenue.

Early access clauses are a common feature. This allows the owner to start the fit out of some of the components for the fit out of the racks whilst the contractor progresses to completion. Good definition around early access will benefit both the owner and contractor and manage the expectation and avoid disputes.

Intellectual property clauses are often the norm in data centre construction contracts because there is a tendency for the design and construction methodology to be inventive and in a state of perpetual progression. Often these rights to protect the inventor and owner of this intellectual property have a grant of rights in the form of licences and permits of use.

Warranties are a feature of data centre contracts that may extend beyond the one-year norm in other construction projects. Regardless of the statutory provisions, it is often the case that electrical and mechanical active equipment has a minimum warranty of two years stated in the contract, with 10 years for water tightness of the roof and façade and components touching the soil. Civil works may carry a warranty of five years. All of these are considerations for the owner and must be priced by the contractor. Having an equipment warranty can assist with after-sales services and reduce replacement costs.

Performance bonds and guarantees can provide specific financial protection if reduced efficiency impacts the bottom line. This may be affected by factors including design issues and operating environment and mitigate against improper installation/construction.

A further tender process will usually be held to appoint the **equipment suppliers/subcontractors**. The data centre customer may have a supplier preference for some of the specialist equipment such as the generators, UPS, cooling, and switchgear and therefore make these named or nominated suppliers within the design and build contract, for example, to keep a common type of equipment/ supplier throughout different phases of a facility. Many of the hyperscale data centre customers have framework agreements with specialist suppliers that already have price, maintenance, and warranty agreements. Therefore, whilst the contractor has many designs and build obligations, they may be constrained in design and the suppliers chosen by some clients. After-sales support and service in a given location should be considered during procurement. This can lead to operational risk where there is limited access to expertise and parts. See Chapter 7.

Plant and equipment and its associated shipping, installation, and commissioning are split into packages. It is important to ensure there are no scope gaps. For example, generators may be in one package, but cabling is within the scope of another subcontractor.

To ensure the equipment proposed is in line with the specification, the contractor will issue **technical submissions** with details of the make and model of equipment they propose to supply for approval by the client/designer.

The contractor may raise **Requests for Information** (RFIs) where there are ambiguities or gaps in the specification or what has been specified is not available. The designer's formal response is then tracked in the project documentation.

Data centre procurement usually crosses borders. The use of supply contracts on international projects is made possible through the service of options relating to changes in the law and multiple currencies. Additionally, these contracts cater for projects requiring a lengthy design and fabrication process and options, allowing for a price adjustment for inflation, advance payment provisions, and corresponding advance payment bond provisions and project bank accounts.

Additional considerations from a logistical and contractual perspective arise where contractors or employers are required to trade with overseas manufacturers. Such concerns increase where the overseas manufacturer is situated outside the European Union or similar domain, notably in tax, import and export licences, customs issues, jurisdictional considerations, and enforcement concerns.

Legal and practical drivers determine the contractual approach to be taken in international contracts. The legal concerns include which governing law is to apply, the forum and venue for dispute resolution, and, depending on the governing law, adjusting the contract to consider the potential misapplication of implied terms and statutory provisions to which either or both the parties are accustomed. The key elements of practical risk are the arrangements for the transportation of goods, the complications due to more extended delivery periods such as payment arrangements, cash flow concerns, finance and security arrangements, additional insurance requirements, transfer of title issues, and currency fluctuations.

It is often the case in construction or engineering projects where goods or materials are required to be sourced from overseas, that such goods are critical or require a lengthy design or fabrication period. Depending on the programme, this may make such items long-lead items, requiring payment to be made before delivery. In such circumstances, the parties will need to consider security arrangements, inspection provisions, and potential audit rights to check on the utilisation of funds and the progress of the fabrication of the goods.

Even where goods do not require a long manufacturing period, the seller will wish to be paid on dispatch of the goods and not on delivery to the buyer and often use **Bills of Lading**. Bills of lading are used to shorten the time the seller will have to wait for payment and still provide comfort to the buyer before the delivery of the goods. Because of the potential shipping period and customs clearance, the seller's position is understandable. A bill of lading accomplishes several functions:

- The terms of carriage of the goods are contained within the bill.
- It acts as a receipt for the goods to the seller and acts as evidence of the good conditions of the goods at loading.
- It evidences the title of the goods. Different types of bills of lading exist depending on the nature of transport of goods. The typical requirements include a shipping bill of lading, airway bill, commercial invoice, certificate of origin, packing list, and certificate of inspection.

The responsibility for obtaining export licences will typically fall on the supplier, mainly where the manufacturer is well-established and proficient in international trade. The obligation needs to be set out contractually. Suppose it is critical to the project that the licence, consent, permit, or exemption is obtained. In that case, the purchaser may wish to ensure it is granted before the main construction contract or at least the construction program becomes effective, irrespective of whether the purchaser of the goods is the employer or contractor of the project.

In general terms, the choices of contractual governing law and jurisdiction made by the parties are usually applied subject to the intervention of necessary rules of law. Jurisdiction is concerned with which courts or arbitral tribunals will hear the dispute. Governing law determines which law applies to the contract. In arbitration proceedings, the rules governing the procedure of arbitration, as distinct from the contract in dispute, are the law of the country where the arbitration is held.

Construction

In the United Kingdom, health and safety are covered by the Construction Design and Management (CDM) regulations.[14]

Construction Design and Management 2015 (United Kingdom)

Commercial clients have specific duties under CDM:

- make suitable arrangements for managing their project, enabling those carrying it out to manage health and safety risks in a proportionate way. These arrangements include:
  - appointing the contractors and designers to the project (including the principal designer and principal contractor on projects involving more than one contractor) while making sure they have the skills, knowledge, experience, and organisational capability
  - allowing sufficient time and resources for each stage of the project
  - making sure that any principal designer and principal contractor appointed carry out their duties in managing the project
  - making sure suitable welfare facilities are provided for the duration of the construction work
- maintain and review the management arrangements for the duration of the project
- provide pre-construction information to every designer and contractor either bidding for the work or already appointed to the project

---

14 UK Health & Safety Executive (2015). The Construction (Design and Management) Regulations. UK Health & Safety Executive.

- ensure that the principal contractor or contractor (for single-contractor projects) prepares a construction phase plan before that phase begins
- ensure that the principal designer prepares a health and safety file for the project and that it is revised as necessary and made available to anyone who needs it for subsequent work at the site

For notifiable projects (where planned construction work will last longer than 30 working days and involves more than 20 workers at any one time; or where the work exceeds 500 individual worker days), commercial clients must:

- notify HSE in writing with details of the project
- ensure a copy of the notification is displayed in the construction site office

Contract preliminaries may be defined by the client to specify general conditions and requirements for the execution of the works, covering items including health and safety, design management, programme, risk management, organisation, reporting, site logistics, environmental management, quality management, cost control, change management, and document management. The site will be set up and managed by the general contractor who provides facilities for the visiting subcontractors. Health and safety provisions include site induction for all visitors, use of PPE (personal protective equipment), risk assessments and method statements (RAMSs), and permits to work (PTW).

Elements of the detailed design may come under the contractor design portion (CDP). The contractor may identify buildability issues with the design and propose alternatives. It is important that it is clear where design responsibility lies.

The contractor is responsible for creating a programme which coordinates the various elements of work. It is important that the sequence has been carefully considered and the task durations properly estimated, with some degree of float in order to build a realistic programme. It is likely that as the project gets underway, there will be changes to elements of the programme; unforeseen issues may come up such as late delivery of components, installer availability, or failed equipment which needs to be replaced. Progress is tracked during the project by reporting on how close the reality on site matches the programme. The contract may include financial penalties for delayed handover.

Depending on the site, there may be civil engineering groundworks to complete before building construction can commence; this phase may be the most difficult due to uncertainty. A key milestone in the construction phase is weatherproofing the building. Following this, the fit-out of internal services can start.

A process to ensure installation quality is important. This involves supervision of subcontractor work, benchmarking of installations, and snagging, whereby inspections identify defects which are tracked to close out.

Commissioning

This is the process of setting up and testing equipment and systems and is typically split into five levels:

Level	Description
1	Factory acceptance testing (FAT) or factory witness testing (FWT)
2	Installation checks and dead testing
3	Energisation and start-up (live testing)
4	Systems testing (or FPT = Functional performance testing)
5	Integrated systems testing (IST)

An independent **Commissioning Manager** may be appointed by the main contractor to lead the commissioning process, whilst the client may engage with a **Commissioning Agent/Validation Engineer** to ensure the commissioning process demonstrates that the installation performs in line with the specifications. In order to determine how to commission the installation, the commissioning team has to understand the design intent. Early engagement allows the commissioning team to identify any commissionability issues, e.g. measuring points missing; they may also identify design issues and potential risks. This can be particularly important for multi-phase projects where connecting onto a live facility or where future phases will do this.

Testing and commissioning requirements including pass/fail criteria may be specified by the designer and developed as the project progresses by the contractor, their commissioning manager and the client's commissioning agent.

Level 1 commissioning usually applies to key items of the plant such as generators, UPS, chillers, and CRAC/CRAH units. Key stakeholders from the project team visit the factory and witness a pre-agreed series of tests delivered by the supplier. The purpose is to demonstrate that the equipment meets the performance requirements before it is shipped to the site for installation. For cooling equipment, this usually involves testing the equipment in an environmental chamber which replicates the design conditions, typically high summer temperatures (site testing will occur with whatever the ambient temperature is at the time).

Level 2 commissioning covers install checks and everything which is needed prior to energisation. This includes electrical dead testing, pressure testing and flushing of pipework, and BMS point-to-point tests.

Level 3 commissioning is the process of setting up equipment once power is available – power on is an important milestone in the project programme. This includes checking phase rotation, equipment settings, operating equipment in different modes under load, and BMS point-to-graphic checks.

Level 4 commissioning involves systems testing, i.e. after all of the equipment within the system has been commissioned, testing the operation of the system as a whole.

Level 5 commissioning or IST checks that the interfaces between systems operate correctly. Typically, this is a mains fail test or series of mains fail tests, which demonstrate that in this scenario, the UPS supports the critical load, the generators start and then take up the load, the cooling systems automatically restart and temperatures within the critical environment are maintained within acceptable limits (see Chapter 5 for the description of thermal runaway). Successful IST is usually a requirement for Practical Completion (PC).

The client team usually witnesses a sample of tests at Level 2-3 and all of the tests at Level 4–5. The contractor should have satisfied themselves that systems are ready for witnessing before offering them to the client team; this usually means that the test has already been successfully completed in a dry run and avoids surprises and delays. Typically, there are checkpoints for sign-off between each stage in order to avoid progressing to the next stage of testing without being ready. This may be described as authorisation to proceed (ATP) or tagging (equipment achieves a particular tag after each level of commissioning). Issues which are raised during commissioning may be tracked on an Issue Resolution Log (IRL). Documenting the commissioning results provides an auditable record of testing which can be used as a reference for the operational team after the handover. It also provides protection for the designer and contractor where issues arise after handover – they can refer back to test records to demonstrate that on a given date, the systems performed as required.

It is not possible to test every combination of operating scenarios. The commissioning process should include the key ones to identify any issues and allow them to be rectified and retested. In addition to programme and budget constraints, other limitations of commissioning include ambient conditions during testing, restrictions arising from interfaces with live systems (e.g. later phases of multi-phase build) and available test equipment (e.g. IT load is typically simulated using heaters but these usually have different airflow and temperature characteristics to IT equipment).

In addition to the stakeholders directly involved in the commissioning process, the operations team benefits from being involved as part of their training on the systems that they will be responsible for after the handover. This is a unique opportunity to learn how systems behave and the required procedures in failure scenarios which do not occur often, as well as gain familiarity with normal operating modes. In addition, product-specific training is usually provided by the suppliers.

As commissioning occurs at the end of the project, there can often be pressure to reduce the time available to avoid delays to project completion, particularly where there have been earlier delays. Taking shortcuts in commissioning usually leads to delays at later stages of testing or hidden faults which do not manifest until the facility is live when they have more impact and are more costly to rectify; this reintroduces the risk that the commissioning process is designed to address. Commissioning helps identify and close the performance gap between design and operation.

Handover (see also Chapter 7)

The contract administrator certifies PC when all the works described in the contract have been carried out. PC is the date on which a project is handed over to the client. The term 'substantial completion' may be used in some forms of contract, particularly in the United States.

At PC, a number of contractual mechanisms are enacted:

- Releasing half of the retention (an amount retained from payments due to the contractor to ensure they complete the work). Historically, this was 5, 2.5% on PC, 2.5% on making good defects 12 months later. Most contracts are now 3, 1.5% on PC, and the remaining 1.5% after 12 months when all defects have been addressed.
- Ending the contractor's liability for liquidated damages (damages that become payable to the client in the event that there is a breach of contract by the contractor – generally by failing to complete the works by the completion date).
- Signifying the beginning of the defect liability period.

As the client now takes possession of the works for operation, the project is typically required to provide a series of documentation at this stage, including operating and maintenance manuals, record drawings, and health and safety files.

Some projects may include milestones such as 'sectional completion' or 'beneficial use,' where the client has a partial handover of the project prior to PC.

The use of the soft landing framework[15] adds a level of enhanced 'aftercare' to help optimise building performance.

Projects may include a lessons-learned workshop after the handover where stakeholders can share their reflections on what worked well and areas for improvement. Issues raised may relate to processes, resources, or design, for example and can help participants identify things to do differently in future projects.

---

15 Agha-Hossein, M. (2018). Soft Landings Framework 2018: Six Phases for Better Buildings. Bracknell: BSRIA.

Operation

The **Operation and Maintenance** or **Facilities Management (FM) team** may work directly for the client or be an outsourced subcontractor. They are responsible for day-to-day operations of the facility after handover.

The power and cooling equipment may be maintained by the FM team, the supplier, or another specialist; these subcontractors are typically engaged via the FM team.

During the operational phase, smaller projects may be required to upgrade or replace infrastructure.

One of the activities required during operation is capacity management; tracking actual power/cooling/space/network consumption versus capacity. This along with future growth forecasts helps predict when the facility will be full and trigger the requirement to plan for how to accommodate future capacity, for example, fitting out future data halls, building a new facility, and purchasing colo space.

Feedback from the operations team's experience of operating the facility should be sought so that improvements can be implemented in future projects.

Either within the same team or under a separate department, IT professionals maintain the IT infrastructure and networks, including physically installing/decommissioning hardware and network cabling – IT Ops. There may also be client service/remote hand teams where this is done on behalf of the end user.

7

Operational Issues

Handover

Once the data centre has been designed, constructed, and handed over, the project phase finishes and the facility starts operation. Handover is a critical milestone in the facility's lifetime, where the knowledge from the project team is shared with the operations team. Most of the stakeholders from the project phase will no longer be involved with the facility unless they have been engaged in a soft landing[1] or other post-project activities. As part of the handover process, the maintenance team which is now responsible for the operation of the facility should have been involved in the commissioning phase and undergone training to understand how the systems work in different scenarios. This is particularly important where the design is complex/unusual. This unique opportunity requires resourcing:

1) If the maintenance team is outsourced, they need to be engaged in advance of the handover.
2) Where the new facility is an extension of scope, the maintenance team still need to be able to do their day job, i.e. maintaining existing facilities. Some organisations use a Transition Manager to support the maintenance team during this phase of the project.

The handover process also involves a transfer of record documentation to the operations team, which includes operation and maintenance manuals (including commissioning results) and record drawings. Unfortunately, the quality of the information shared is often lacking. This may be due in part to time pressure at

---

1 BSRIA (2018). *Soft landings framework 2018 (BG 54/2018).*

the end of the project and the fact that it is a low priority for the project team. Understanding how the facility is designed to operate is important to allow the operations team to ensure its performance is optimised. The performance gap between the design and operation of buildings is well documented.[2] Although there are contractual boundaries between project parties, a collaborative approach which maintains a focus on the operational phase can help close this gap. Where a complex design is not understood or does not operate as expected, it is likely that it will be set to operate in a manual mode (predictable but with limited functionality) and may also be less efficient and have a higher operating cost.

Legacy Facilities

Sometimes an operator may take on an existing facility, for example, through acquisition. In these circumstances, where the facility has been built for another, there may be some elements of the design/installation which do not fit the new user's requirements. Commissioning and record documentation may be lacking which also presents a risk; there may be faults with the installation which have not yet manifested. As part of their due diligence review, the new owner may wish to undertake a single point of failure (SPOF) review to understand risks associated with the facility; another tool is FMECA (failure modes, effects, and criticality analysis). Known risks may be recorded on a risk register and depending on their magnitude, the business may choose to invest to remove or mitigate them.

Operations Team

The operations team is also known as 'Ops,' the maintenance team, facilities management (FM), or similar. In most data centres (size/criticality dependent), there is a 24/7 presence for monitoring and maintenance, covered by shifts, usually in addition to a wider team working in office hours. This is to ensure that should there be an issue affecting the critical infrastructure, this can be identified and dealt with promptly, causing minimum disruption to service. The maintenance team usually monitors system statuses from the NOC via the BMS head end on-site, and sometimes this is also visible remotely. The BMS (see Chapter 5) is typically set up to identify alarms with different priority levels, for example, cooling unit filter needs changing = low priority, temperature in the data hall exceeds

---

2 Operational performance of buildings: CIBSE TM61: 2020. London: The Chartered Institution of Building Services Engineers.

SLA = high priority. As well as being identified at the head end, high-priority alarms may also be shared via email/text message to key team members. When there is an event such as a mains failure, the system may become flooded with a large number of alarms which can be difficult to navigate, rather than assisting the operator to identify the issue(s) and required action. Priority levels of alarms should be considered with this in mind. Some facilities have a service desk where clients can make contact via phone/email and raise incident tickets.

In addition to monitoring via the BMS, it is often a requirement that each shift undertakes a walkaround to monitor plant statuses in person and identify any anomalies (not all issues are necessarily picked up by the BMS and this can also fail). At the shift handover, the outgoing shift will share any known issues or planned activities with the new one starting.

The maintenance team may be in-house staff working directly for the data centre operator or outsourced. Important elements to get right include:

1) Resources
2) Processes and documentation
3) Communication
4) Training and development
5) Working environment

Outsourced maintenance teams are often more popular with less mature data centre operators looking to bring in external experience and processes. High-performing outsourced contracts tend to have a 'one team' culture, rather than 'us and them'. The maintenance contract may be tendered every five years or so. Where the contract provider changes, it is common for at least some of the site staff to be retained and continue working for the new provider.

Contracts vary but there are typically service-level agreements (SLAs) between the outsourced maintenance provider and the data centre operator and/or between the data centre operator and their client(s). The service provider may have to report on their performance, including compliance with SLAs and other key performance indicators (KPIs), and may be subject to financial penalties where they underperform. It is important that the contract reflects the client's expectations and any SLAs agreed with their tenants. There are cases where there is a disconnect between the sales team and the design reality; what has been sold is not aligned with the facility's performance, putting pressure on Ops to deliver beyond the design limitations, or having to deliver a requirement which is no longer best practice (e.g. a very narrow temperature and humidity range). SLAs may include response times for critical incidents, requirements for power availability (target zero unplanned outages to IT power supplies), and data hall environmental conditions, e.g. average temperature and humidity across specific sensors must remain within the specified range. It is important to consider the location of the SLA temperature sensors, e.g. avoid locating them in the hot aisle which is supposed to be hot. Also consider whether a temperature SLA breach has a time dimension, i.e. instantaneous or average over a period.

Procurement of maintenance contracts is heavily cost-driven. The maintenance contract is typically set up so that the provider earns a markup on subcontractor works. This does not necessarily promote quality of service nor incentivises providers to look for underlying causes of failure – it is more profitable to continuously replace failed components. The maintenance team also may not have the expertise to identify all causes of failure, e.g. design issues.

Whichever model is used, it can be challenging to find and retain staff with the right skills and experience. New team members may not have data centre experience and thus a specific induction and training process is required to make them aware of the challenges and risks of working in a critical environment and the facility-specific requirements.

The data centre is not a static environment and installation, decommissioning, configuration changes, upgrading, and maintenance of IT equipment are usually ongoing via the client services or 'remote hands' team. Many IT changes can be undertaken remotely but some physical interventions are still required. The maintenance team may provide support by installing and energising rack power supplies.

Where workers are unionised, jurisdiction-specific protocols apply for stakeholder engagement.

Uptime and Failures

Components will fail. However, most facilities are designed in a way which includes some level of redundancy to plant/systems. This means that one failure should not cause a service outage such as an interruption of power or loss of cooling.

Most failures are due to human error or unawareness and typically, a combination of root causes. Note that human error does not just include an operator making a mistake on the day but design and management issues including processes and resources – these tend to be more significant than individual operator errors. Analysis of failures from a broad range of industries has found that a universal learning curve applies, whereby as experience increases, the failure rate decreases for both organisations and individuals.[3] The risk of failure can be

---

3 Duffey, R. and Saull, J. (2008). *Managing Risk: The Human Element.* John Wiley & Sons.

reduced for a facility which effectively addresses its operational vulnerabilities, staffed by individuals with the right knowledge, experience, and attitude.[4] It is important to create an environment where mistakes are learned from, rather than a blame culture, for example, through undertaking root cause analysis (RCA) to identify what caused a failure and how this may be avoided in future. Some organisations punish failure by firing staff in the belief that this will fix the problem. However, inexperienced staff are more likely to cause failure and high staff turnover represents a business risk. Attributing the blame to an individual/individuals can deflect blame away from others in the organisation which also had a contributing role and mean that management issues remain unchallenged.

There is an operating philosophy that favours automation over human intervention in order to avoid the issues associated with humans. There is often scope to streamline some of the manual tasks undertaken by operators. Greater automation also has its challenges including upfront cost and complexity. There is still a human interface – with the designer, installer, and testing team, and when things go wrong or an unanticipated event happens, the number of people with the know-how to respond is limited.

Recovering from a failure contains several elements as illustrated in Figure 7.1.

The first step is recognition of the issue, typically through monitoring, e.g. via BMS. It is important that the operator is aware of the issue prior to their client(s), who will make contact if a service outage is experienced.

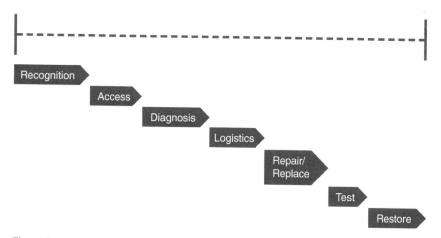

Figure 7.1 Duration of downtime.

---

4 Cameron, D., Tozer, R. and Flucker, S. (2016). *Reducing Data Centre Risk and Energy Consumption Through Stakeholder Collaboration*. Edinburgh, UK: CIBSE Technical Symposium.

Access is required in order to make a diagnosis (a BMS alarm may provide sufficient detail, although this is verified locally). Some equipment may be able to be remotely monitored by the manufacturer. However, most data centres consider external network access to their equipment too risky.

Diagnosis may require specialist attendance. Often the maintenance team will have a call-out SLA with the provider, for example, they will be on site to investigate the issue within four hours.

Logistics includes, for example, obtaining any required spare parts. The project usually provides or at least identifies critical spares – items which are recommended to be stored on site so that they are immediately available for replacement when required.

Accessibility of local expertise has an impact on repair times, e.g. if the nearest available qualified engineer is not based in the country. This should be considered during the procurement of equipment.

Then the repair/replacement can take place, which requires testing before being put back into service.

Not all sites are manned which can increase timescales.

Not all failures result in service interruptions. However, should a power and cooling failure result in a service interruption for IT, the total downtime includes the time to restore IT systems. A power interruption of one second to the critical load could result in several hours of IT service interruption[5] as illustrated in Figure 7.2.

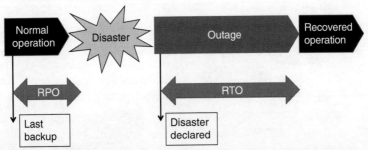

- TTRS = MTTR (M&E) + time to validate IT
- TTRS is the customer downtime

RPO = Recovery Point Objective = the most recent point in time to which system state can be recovered onto backup system
RTO = Recovery Time Objective = the target time between when disaster is declared and when service is recovered on backup site

Figure 7.2 Time to Restore Service (TTRS).

---

5 Bauer, E., Adams, R. and Eustace, D. (2011). *Beyond Redundancy*. John Wiley & Sons.

Maintenance Processes and Procedures

Local code may mandate numerous statutory requirements for maintenance, with a safety/efficiency focus.[6]

Maintenance of critical facilities usually goes beyond this to help ensure the reliability of systems and equipment. A maintenance management system is used to programme the required maintenance for all assets so that this can be scheduled by the ops team. There are different maintenance philosophies; most facilities operate a programme of planned preventative maintenance (PPM) which is the simplest method to set up. This is where components are replaced on a regular basis regardless of whether required or not. The idea is that carrying out maintenance of assets in line with the manufacturer's recommendations can maintain reliability, reduce reactive maintenance, i.e. break-fix, prolong the operational performance and operating life of the equipment and reduce the total cost of ownership. However, it may increase TCO, as some components will be replaced before they require it and some will fail before replacement, potentially resulting in additional costs depending on the impact of failure.

One alternative is reliability centred maintenance (RCM) where the reliability and criticality of each component are analysed and maintenance tailored to suit. For example, a light bulb in a non-critical area might not be replaced until it fails but light fittings in a critical switchroom would be checked on a regular basis. Predictive maintenance is where critical components are monitored to highlight problems prior to failure, e.g. bearing noise.

Maintenance work is typically covered by Maintenance Operating Procedures (MOPs) or Standard Operating Procedures (SOPs) which document the approved methodology for undertaking specific activities, e.g. isolating a chiller for maintenance. Similar to during the construction phase, health and safety controls will be in place such as permit to work, use of electrical switching plans by authorised persons (APs) and senior authorised persons (SAPs) for HV installations and lock out tag out (LOTO) procedures.

Often the maintenance is not undertaken by the maintenance team directly but by specialist subcontractors who they manage, often the equipment manufacturers or suppliers. Regular supplier performance management reviews may be undertaken to ensure service quality and highlight areas for improvement.

In order to maintain an auditable record of activities, there is a certain amount of documentation associated with scheduling and closing out maintenance activities. Checklists can prevent issues such as the generators being left in

---

6 Bleicher, D. and Blake, N. (2022). *Statutory Compliance Inspection Checklist (BG80/2022)*. BSRIA.

manual after their maintenance (which would prevent them from starting automatically in a mains failure).

Typical maintenance activities include generator testing on load, to help ensure that when called upon to run in a real mains failure scenario, the system operates as designed. Some facilities may have an in-built load bank for this purpose or hire a temporary load bank (facility load may be low after handover and beyond). Some operators test generators off-load due to the perceived risk of testing on load. An off-load test is limited in its scope and does not fully test the system. Better to test on load under controlled conditions with the right people on site to recover if needed, rather than avoid testing it and finding a problem during a real failure scenario.

Poor water quality in chilled water/process water systems may result in corrosion, calcification, and bacterial growth (including pseudomonas and legionella) which cause operational issues including component failures, loss of capacity, and health and safety risks. Water treatment[7] and valve exercising are some of the activities which may form part of the management regime. In the United Kingdom, ACoP L8 covers the control of legionella bacteria in water systems.[8]

Emergency operating procedures (EOPs) cover what to do in unplanned situations, for example, when there is a mains failure and the generators do not start – there may be a manual process to try and start them. The objective is to try and prevent downtime or recover from an outage and avoid people rushing in panic, making mistakes and making the situation worse. The risk of human error is not just with inexperienced staff doing the wrong thing due to unawareness; there is a risk of complacency with more experienced operators who may skip some of the checks required when undertaking a task. Scenario training with dry runs of the EOPs can be useful to ensure the site team are ready to respond.

Alongside EOPs are alarm response procedures (ARPs), which identify steps to take following an alarm – this may include not only which EOP is applicable but how to communicate what is happening with other stakeholders, e.g. escalation process.

It is impossible to cover every emergency scenario which might occur; the most commonly anticipated ones should be prepared for. One scenario which may be covered by an EOP is overheating in the data hall. This might identify steps to determine the cause of overheating such as cooling plant failure. However, EOPs typically do not anticipate design/installation issues, for example, a heat rejection

---

7 Water Treatment for Closed Heating and Cooling Systems (BG 50/2021) BSRIA.

8 UK Health & Safety Executive (2013). Legionnaires' Disease: The Control of Legionella Bacteria in Water Systems. Bootle: Health And Safety Executive. https://www.hse.gov.uk/pubns/books/l8.htm.

plant is installed too close together creating a local heat island effect which downrates the cooling capacity and causes refrigeration circuits to trip on high pressure. The problem might have been less apparent during commissioning if this was undertaken during cooler ambient conditions. Some improvised solutions may be employed in this situation such as spraying water over the heat rejection plant to cool it down (this can cause other issues).

Managing Change

There is also risk associated with maintenance, particularly when this requires the system to be operated in a state of reduced resilience, e.g. N rather than $N + 1$ (see Chapter 5). For this reason, it may be scheduled at a time of reduced business risk, e.g. evenings or weekends.

Any intervention in a critical system has risk associated and thus there are procedures in place to control how works are undertaken. Depending on the assessed risk of the works, it may be necessary to submit a change request to the business in advance. The date and details of the proposed procedure to undertake the works will be reviewed and approved/rejected. It may have to be rescheduled at a less risky time.

The risk profile of changes in a live facility is likely to be higher than during the project phase due to the potential impact on mission-critical services. For some businesses, the risk of business impact during certain time periods may be so great that works are not permitted during a 'change freeze'.

Where maintenance is continually delayed, this also introduces risk. Better to undertake the maintenance, identify and rectify any issues when the right team is present, rather than have the equipment fail at an unplanned time.

Changes include not only maintenance but plant replacements and upgrades during the facility's lifetime. Guidance is available on asset lifetime; replacement is usually considered when an asset is no longer economic to repair, becomes obsolete, is deemed too inefficient, is subject to legislative action, or is otherwise overtaken by an alternative solution.[9] Obsolescence may apply to software or hardware; the product may no longer be supported for service or parts are no longer available.

Small projects, e.g. equipment end-of-life replacement may be undertaken by the ops team rather than a separate project team. Similar to during a building project, communicating the up-to-date status of the system, maintaining an

---

9 Harris, J., Craig, B. and Chartered Institution Of Building Services Engineers (2014). *Maintenance Engineering and Management: A Guide for Designers, Maintainers, Building Owners and Operators, and Facilities Managers: CIBSE Guide M*. London: Chartered Institution Of Building Services Engineers.

auditable record of the works and updating record information are essential for the team managing the facility.

Capacity Management

It is important that the operator has an understanding of the design capacity of the facility and where the bottlenecks are. For example, the electrical capacity may exceed cooling capacity. Therefore, the cooling capacity is the limitation. There may be areas of the data hall which support high-density cooling and others which do not. If the UPS system has a $2N$ design, under normal operation, the maximum load is 50% per stream (see Chapter 5). A bespoke load tracking tool may be employed in order to monitor usage. This could be a spreadsheet or some kind of data centre infrastructure management (DCIM) platform. Load growth monitoring helps predict when additional capacity may be required and up-to-date load information should be used when planning maintenance.

Similarly, it is important to monitor how much compute, storage, and network capacity is available. Hardware asset management is an important process to avoid the operation of 'zombie servers,' which are powered on and using resources but where they are no longer required and their original owner has not reassigned or decommissioned them.

One challenge is predicting future hardware trends, i.e. what is coming and when (depends on business demand and products available) and how much power will it really use (normally significantly less than the published nameplate value). When does additional capacity (self-built or outsourced) need to be available and are the timeframes compatible with demand?

Training

Training is an important investment to help prevent some of the issues discussed in this chapter. This includes not only technical fundamentals but also site-specific applied knowledge and non-technical skills, e.g. management. There is a widely reported skills shortage at all levels in the industry. The workforce is ageing and growing faster than new joiners can be recruited and gain experience. This is not helped by a lack of diversity in the sector. The Uptime Institute's 2022 annual survey reports that more than three-quarters of operators have a data centre workforce of around 10% women or less.[10] The shortage of staff means that there

---

10 Pearl, T. (n.d.). *Uptime institute global data center survey results 2022*. Available at: https://uptimeinstitute.com/resources/research-and-reports/uptime-institute-global-data-center-survey-results-2022 [Accessed 4 Oct. 2022].

is high demand and individuals may move positions due to competitive rates of pay in other facilities/organisations.

Not all organisations prioritise training. However, there is a significant risk when operating with untrained staff. Training does not just apply to inexperienced staff, it is an important pillar for continuous improvement. Regular scenario training, where operators practise how they would react to a given incident is a useful drill to help prepare for a real-life incident.

Performance Optimisation – Beyond Reactive Maintenance

The role of the maintenance team should not just be to react to and prevent failures but also to ensure that systems are working in an optimised way. This can result in reduced equipment failures, maintenance costs, and increased equipment lifetimes; also reduced energy consumption and cost. To do this requires the right skills and resources to analyse and act on performance data. For example, the facility is operating at part load and so is not operating at the design PUE. In many cases, it is accepted that the design does not cater for (low) part loads and no action is taken to make improvements. However, there are typically opportunities to adjust the system control parameters to better suit the facility load without redesigning the facility. These can be implemented with minimum risk and result in important energy savings. There are several barriers to change, e.g. ops team lack skills and confidence to challenge design (they are not designers), lack of resources, perceived risk, and potential benefits not well understood (business case not well presented).

8

Economics, Investment, and Procurement of Data Centres

During the COVID-19 pandemic, data centres became one of the best performers for real estate investors because of the increased demand for Web services.

The underlying need[1] was driven by many factors, including the digital transformation of adopting digital technology to transform services or businesses by replacing non-digital or manual processes with digital processes or replacing older digital technology with newer digital technology requires reliable data centre capacity that can scale quickly and meet demand.

Therefore, the growth of investment in and procurement of data centre assets, both brownfield and greenfield, looks to continue. Data centres are repeatedly deemed[2] an attractive target for investors looking for a stable and dependable source of recurring revenue, as hyper-scale operators such as AWS, Google, IBM, and Microsoft tend to sign up for multi-year contracts. It is also a scalable investment meaning that once a business footprint has been established, the organic growth of the sector will permit incremental opportunities. The global data volume is forecast to be exponential as (Figure 8.1):

To put this graphic of exponential growth, 1 Zettabyte is approximately 1 billion Terabytes. An average laptop has c. 256 GB of storage, c. 3900 million laptops per Zettabyte. Perhaps a nebulous calculation may illuminate the scale of the computing power in these data centres from a global perspective.

Various types of data centres facilitate this demand, each with unique investment opportunities.

---

1 See Chapter 2 of this book.

2 DLA Piper (n.d.). *Global investment in data centres more than doubled in 2021 with a similar trajectory this year / News / DLA Piper Global Law Firm.* [online] Available at: https://www.dlapiper.com/fr/france/news/2022/06/global-investment-in-data-centres-more-than-doubled-in-2021-with-similar-trajectory-this-year/ [Accessed 01 May 2023].

Data Centre Essentials: Design, Construction, and Operation of Data Centres for the Non-expert, First Edition. Vincent Fogarty and Sophia Flucker.
© 2023 John Wiley & Sons Ltd. Published 2023 by John Wiley & Sons Ltd.

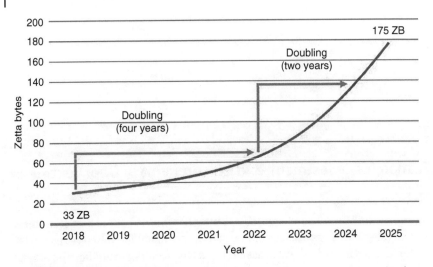

Figure 8.1 Growth of data centre capacity.

Enterprise

Generally, enterprise centres are custom data centre operations owned and operated by a single corporate owner for the owner's particular needs and use. They are not typically offered to third parties and are designed to house the corporate data backups and information technology (IT) functions used in that business's unique daily operations.

Typical businesses such as banks, supermarkets, and government departments have their own enterprise data centres. These enterprise data centre operators may themselves build their data centre facilities or locate their data centres within colocation facilities, with many companies doing both. Some companies not only operate their own enterprise data centres but also sell some colocation space to other companies, and this is a logical option for enterprise operators who find themselves with spare capacity.

Enterprise operators may build their own data centres, but mostly it makes commercial sense to use a third-party provider such as a development partner. Once an organisation's data requirements reach a specific size or become mission-critical where service disruption has significant adverse consequences that may generate liabilities, this data will need to be housed in an environment with guaranteed security, continuity of power supply, and connectivity. Companies have several options. They could build their own facility or lease space from a wholesale or a colocation operator and still manage their IT themselves. They could outsource the whole IT function to an IT services provider who, in turn, may have their own data centre or have taken space within a colocation space. They could even buy and develop a site and then contract a third party to manage their IT for them within it. The procurement of IT space is full of options that involve fundamental economic considerations.

Many businesses do not plan to own or operate all of their facilities. The options are to shift workloads to cloud providers or use colocation facilities. It is simply because data centres are not the core business; however, they may retain highly classified data within their domain. These hybrid solutions are where a small proportion of data is secured within the enterprise, and the remainder is outsourced.

Many firms prefer to deploy infrastructure into one or many colocation facilities in order to maintain commercial tension and avoid a single point of vendor failure. Then it may interconnect with other enterprises and service providers housed in the same facilities through cross-connects.[3] Retail and wholesale are the two most convenient routes to colocation. The needs of the firm defined the best colocation arrangement. In general, retail allows more flexibility, but wholesale offers more control.

Colocation

This data centre primarily serves as a platform to lease space or share services with other users. It is like a data hotel where tenants can lease as much or as little space or support as they need. Colocation data centres come in all shapes and sizes, having lots of tenants or just one. The colocation business model is to lease out and provide space and services to a third party for a monthly fee or rent.

Retail Colocation

For companies needing flexibility with changing IT requirements, retail colocation makes sense. Companies may rent a rack, cage, or cabinet space for implementing their own IT equipment in retail colocation. Companies may have little control over the space in this model. However, they do have quick access to fibre cabling, equipment racks, power for rack equipment, cooling, fire suppression systems, physical security, and ancillary amenities.

---

3 volico (2020). What Does 'Cross Connect' Means in Data Centers? | Volico Data Centers. [online] Miami and Broward Colocation | Volico Data Centers. Available at: https://www.volico.com/what-does-cross-connect-means-in-data-centers/#:~:text=But%20what%20are%20cross%2Dconnections [Accessed 1 May 2023].

Wholesale Colocation

A company with more predictable needs may require storing vast amounts of data and running significant business applications in the same location for several years. In these circumstances, a wholesale colocation method may be more advantageous because it permits more control at a lower cost.

A wholesale model also permits corporations to design and build their own internal space. It also demands a commitment to lease more significant space and power. These tenant companies will need to bring all their resources to design and build the space, like IT equipment racks, cabinets, power connections, fibre, and other ancillaries. Those with significant content, like media providers, cloud service providers, hosting, IT-managed services, and telecommunications companies, are all familiar wholesale colocation customers.

Many companies also bring together these approaches, leasing wholesale data centre space in locations around the world. Depending on where their business growth creates demands while using retail colocation in areas with limited growth or uncertain evolving needs exists.

Retail versus Hyper-scale Data Centres

Retail Data Centres

Retail data centres may cater to thousands of client companies that get colocation space using either model. Retail colocation space is offered as a monthly service contract, including power and cooling, similar to how managed and cloud services are offered. They provide what is known informally as 'position, power, and ping', which provides the critical infrastructure, including security, resilient electricity supply, broadband connectivity, and an environment in which temperature and humidity are controlled to suit servers. The colocation provider may sell or lease space within those specialised facilities to companies installing and managing their IT equipment. The term colocation derives from the fact that these customers share or 'co-locate' their IT operations in one purpose-built facility with others. These contracts are typically measured in 12, 24, or 36-month increments. Wholesale colocation space prompts a lease requirement, generally with terms from three to twenty-year contracts.

Hyper-Scale Data Centres

A hyper-scale data centre is engineered to meet the technical, operational, and pricing requirements of hyper-scale companies, such as Amazon, Alibaba, Facebook, Google, Salesforce, Microsoft, and a handful of others. These hyper-scalers need vast amounts of space and power to support massive scaling across many thousands of servers for cloud, big data analytics and computation, or storage tasks. Tailored design criteria meeting user-specific needs, such as specific power resilience and cooling redundancy, are also critical requirements.

A hyper-scale does not permit other companies to connect inside its data centres. Instead, it houses a network extension or node in a retail facility, permitting all the companies in the retail facility to connect to the hyper-scaler.

Investment and Procurement

The data centre market tends to focus on retail and wholesale hyper-scale colocation for investment purposes.

Whether as a company you are looking to self-build a data centre or considering colocating your data centre need, there is an economic assessment that is necessary. Cross-functional business needs often influence the procurement of the most suitable type of data centre facility. These cross-functions may consist of financial and facilities professionals that typically work within the chief financial officer's organisation and IT professionals that usually work within the chief information officer or chief technology officer's organisation. In effect, many strands of consideration may require input from many parts of the business.

In response to these business needs, data centres have become attractive to long-term private capital[4] and those seeking to deploy capital in alternate asset classes with infrastructure-like attributes, revenue growth, and steady long-term income. Data centre businesses are an attractive fit for this type of investment class. The aphorism that 'data is the new oil'[5] symbolises the surge in data centre investment.

Investment in Colocation Facilities

These lease contracts tend to be long-term with creditworthy counterparties of the type mentioned earlier.

The aspects for which data centre operators usually accept responsibility include temperature control, uninterruptible power and electricity efficiency provision, and secure telecommunications infrastructure connections. Data centre operators sometimes take responsibility for cybersecurity at certain levels, but most hyper-scalers maintain their cybersecurity controls and risk protection

---

4 Such as private equity and managed infrastructure funds, pension, and sovereign wealth funds.
5 Rics.org (2023). *Infrastructure funding in the age of big data*. Available at: https://www.rics.org/news-insights/wbef/infrastructure-funding-in-the-age-of-big-data [Accessed 1 May. 2023].

themselves. These requirements are often stated as key performance indicators (KPIs) in the lease's service-level agreements (SLAs),[6] which form part of the overall customer contract.

Some hyper-scale colocation is only for a single tenant to the facility, branded in the colocation operator's name. Sometimes, these discrete data centre facilities may be preferred by some tenant types who prefer not to have the presence advertised in a particular location.

In the single-tenant colocation model, the data centre is a purpose-built facility with specialised infrastructure, including cabling connection services, cooling, fire safety, backup generation, and meet-me rooms (MMRs).[7] However, the internal spaces are controlled by a single hyper-scale tenant. In effect, the colocation operator never 'touches' the data but simply provides the critical facilities supporting facilities.

Whilst single tenants are common, most investment is in the provision of multi-tenanted colocation data centre services for retail and hyper-scale colocation customers. The majority of collation need appears to be for the data storage needs of individual enterprise customers rather than providers of cloud services. This is a significant market segment and is populated by the likes of Equinix, Digital Realty Trust, CyrusOne, China Telecom, and others.

Some contracts with single-tenant corporations can be substantial.[8] Private capital tends to prefer investment in hyper-scale colocation operators, given the greater certainty around revenue stream and credit condition of their long-term and diverse customer base. However, recently CyrusOne and CoreSite have been acquired[9] by private equity.

Private equity firms view data centres as one of the best investment grades. Strong fundamentals[10] help allure investors, whilst the investment entails risks not present in other infrastructure assets. Contract terms are shorter in duration

---

6 While a vendor contract focuses on specific duties for both parties, a service-level agreement (SLA) is used to measure the performance and service quality of the vendor. This can either be a stand-alone document or included in the contract. The primary purpose of an SLA is to spell out the level of service that will be provided.

7 Pettit, V. (2020). *What is a meet-me room?* [online] AFL Hyperscale. Available at: https://www.aflhyperscale.com/articles/techsplainers/what-is-a-meet-me-room/ [Accessed 1 May 2023].

8 North American Data Centers. (n.d.). *Reports Archives.* [online] Available at: https://nadatacenters.com/category/reports/ [Accessed 1 May 2023].

9 www.spglobal.com (n.d.). *Data center REITs CyrusOne, CoreSite Realty acquired in all-cash deals.* [online] Available at: https://www.spglobal.com/marketintelligence/en/news-insights/latest-news-headlines/data-center-reits-cyrusone-coresite-realty-acquired-in-all-cash-deals-67699129 [Accessed 1 May 2023].

10 www.jll.co.uk (2023). *Data center outlook year-end 2023.* [online] Available at: https://www.jll.co.uk/en/trends-and-insights/research/data-center-outlook.

than other types of infrastructure, and the relative speed at which they may be constructed means there is a lower threshold to entry for competitors. Because data centre mitigation from one facility to another is so logistically challenging for tenants, the revenue stream to a particular data centre is dependable if the facility's performance is adequate.

Whilst equity investment is quite active in the market, debt investors have more opportunities to deploy funds to existing operators through, for example, refinancing or upsizing existing facilities due to merger and acquisition or expansion of capital expenditure (CAPEX) requirements. Debt investors, unlike equity investors, can take secondary positions in current bank syndicates. The sheer scale of the CAPEX required by many operators means that it is unlikely that one bank is prepared to cover all the CAPEX, hence the need for several banks to provide the necessary capital levels.

The underlying strategy in most of these procurement deals is to pump funding into existing data centre providers so that the acquired company can increase their market share, thus creating a healthy return for investors. It may be expected that the valuations of these types of funding deals and acquisitions to increase over time as data centre services become increasingly essential.

In financial terms, data centres present some unique characteristics and risks for debt financiers and equity investors compared to other asset classes. While they share many qualities with traditional real estate financing, there are significant differences, and financing approaches applicable to long-term infrastructure assets have increasingly been adapted to data centres. Some data centre business models also present themselves as leveraged finance-style funding packages with significant growth CAPEX expansion flexibility. Attractive high leverage supported by forward-looking EBITDA[11] projections and often with detailed accounting modifications based on an expanding book of customer contracts.

Whilst leveraging cash for data centre business expansion via debt is typically sized and secured against contracted net operating income (NOI), given the more refined pricing compared to equity returns, there is little margin for error in financial performance. The stickiness and robustness of cash flows, the accuracy of operational expense (OPEX) costs and any residual refinancing risk at the end of the funding term are critical areas of focus for debt providers and, therefore, will require considerable input from independent data centre consultants at the due diligence stage. Energy-efficient data centres help reduce costs, attract customers, and boost environmental, social, and governance credentials; therefore, it pays to focus on power usage effectiveness (PUE) metrics when weighing up investments.

---

11 Earnings before interest, taxes, depreciation, and amortisation are a measure of a company's overall financial performance.

Power costs are influential in any data centre investment decision, not solely for commercial reasons. PUE is an especially critical metric in establishing the total cost of ownership (TCO) for leasing a colocation data centre. In addition to rent, a colocation tenant is liable for the electricity cost needed to support their IT infrastructure and the overhead associated with the PUE. Greater PUE efficiency consequently equals a lower TCO.

The greater the power demand, the more significant the environmental impact unless a green energy source mitigates it. Many are seeing the transformation of the data centre sector blending to deliver low-carbon solutions that have integrated power generation and waste heat recovery. An excellent example of this type of solution is Meta's Odense facility in Denmark, which has wind turbines that add renewable power to the facility supply grid and then recovers and donates the heat from the data halls to the local domestic district heating network.

However, investors may need to be more careful of categorising data centre financing solely within a single financing model, whether real estate, infrastructure, project, or leveraged finance. Because of some of the unique features of data centres, strict observance of a particular financing approach may encounter obstacles. Financiers must be flexible to embrace certain features of alternative financing models suitable for the particular business and credit profile. For example, a conventional project finance approach to three-party arrangements for key customer contracts will often be inappropriate. Factors such as the different tenors and the relatively high value of customer-installed equipment are considerations. The customer's rights to use the facility usually mean that collateral contract safeguards are as much for the security of the customer as for the financier.

Power Supply Arrangements

An uninterrupted and reliable power supply is critical for data centre operations. A power supply outage in the data centre may immediately increase the service credits payable to the customers and cause irreparable harm to the data centre provider's reputation, even if only for a brief period.

Therefore, during due diligence, investors should ensure the data centre has secured sufficient power supply for its contracted capacity to the customers from a reliable source. Reviewing any power purchase agreement (PPA) between the power supplier and data centre operator is critical to ensure the facilities' uninterrupted power supply and cooling. A PPA may be physical or synthetic.

These trends have prompted an increasing use of corporate PPAs (CPPAs) that cater to the substantial energy needs of data centres. This increases the complexity of financial due diligence on critical PPA contracts compared to 'standard' power supply agreements.

A CPPA is an energy contract for businesses that purchase renewable electricity directly from a producer. It allows the purchase of green energy directly from the source. It will enable both the consumer and producer the flexibility to fix the price you pay for that energy for a specific period. Power may be supplied via a renewable generation facility or advanced storage with the data centre, supported by CPPAs, private wire or other arrangements.

A physical PPA, commonly with a term over 10–15 years, engages with a renewable energy producer to take some or all of the energy generated by its plant with a well-defined amount of power sold at a fixed price per megawatt hour (MWh). In a synthetical PPA structure, no power is physically traded.[12] In its place, the agreement functions with a derivative contract structure where the offtaker[13] and generator agree on a defined *strike price* for power generated by a renewable energy facility.

In both cases, investors ought to also understand the service level credits pertinent to any power outages and thoroughly review the performance history of the data centre, including whether any service level credits have been granted to customers historically.

It may be preferred if there is an indemnification clause in the power supply agreement where the supplier agrees to indemnify the data centre provider for any loss caused by a power outage.

However, in some Asian markets, this is unrealistic, given that the government controls the power supply, resulting in an effective monopoly in these markets. This is the case in Hong Kong where customers have minimal negotiating power with China Light and Power.[14] In Singapore, power costs are unpredictable and have constrained the market in negotiating new power supply arrangements.

Because power costs constitute a substantial part of overall data centre operational costs, investors ought to investigate how power costs are passed to customers and whether any permits are required in respect of the pass-through agreements.

Investors should also determine if there are any minimum obligations in the power supply agreement, for example, where the data centre provider commits to

---

12 DLA Piper (n.d.). *Corporate Power Purchase Agreements (PPAs): What are they? / Insights / DLA Piper Global Law Firm.* [online] Available at: DLA Piper. (n.d.). *Corporate Power Purchase Agreements.* [online] Available at: https://www.dlapiper.com/en/capabilities/industry/energy-and-natural-resources/corporate-power-purchase-agreements [Accessed 1 May 2023].

13 Editor, F.B.F.L.F.T.T.S. is an, finance, writer S. has 20+ years of experience covering personal, Management, W. and Segal, business news L. about our editorial policies. Segal, T. (n.d.). *Offtake agreements: What they mean, and how they work.* [online] Investopedia. Available at: https://www.investopedia.com/terms/o/offtake-agreement.asp.

14 CLP Group (n.d.). *Home.* [online] Available at: https://www.clpgroup.com/en/index.html [Accessed 1 May 2023].

a minimum payment obligation for a specified period, regardless of how much electricity the data centre consumes during that period. This will negatively influence the data centre's valuation if the data centre provider does not have a significant enough demand pipeline and fails to pass on these minimum commitment costs to its customers.

Two mega trends are sure to shape the data centre landscape going forward. These are the 'twin transitions' of prevalent digitalisation and decarbonising.[15] Digitalisation also presents opportunities for lowering carbon emissions, increasing remote working, reducing business travel, and digitalising supply chains. Many data centre operators are corporate leaders in decarbonising data, committing to net zero carbon and 100% renewable energy goals. They articulate these as fundamental priorities that lie at the heart of their respective business plans and long-term visions for future operations.

Other Complexities

Operational complexities, such as physical and data security, will also benefit from expert advice at the due diligence stage.

Several different factors influence data centre investments. Putting technology aside, investors need to assess everything from the relative attractions of competing geographies to energy efficiency and environmental performance. These new factors coming into play are likely to have an increasing influence on the choice of geography. One is the need for energy efficiency. Some geographies perform better than others in this regard. In the Nordics, there is a lot of renewable energy, and the climate is colder, so you can depend more on ambient cooling, which drives energy costs down. The deployment of new submarine fibre cables adds to the region's attractiveness.

Sweden is ranked relatively high even though it is outside the Frankfurt, London, Amsterdam, and Paris (FLAP) markets, the four European cities with the highest data centre rents in Europe. Sweden's growth highlights the advantages of the Nordics region as a location with abundant, low-cost renewable energy available, addressing environmental concerns using free cooling in a cool climate, and providing a reduced PUE ratio than the other regions.

Another consideration is the shift to decentralised edge computing and the trend towards moving data processing closer to end users. The demand for low

---

15 BloombergNEF (2021). *Data centres set to double their power demand in europe, could play critical role in enabling more renewable energy*. [online] Available at: https://about.bnef.com/blog/data-centres-set-to-double-their-power-demand-in-europe-could-play-critical-role-in-enabling-more-renewable-energy/ [Accessed 1 May 2023].

latency rapid response edge computing using micro data centres complements rather than replaces existing data centre infrastructure. This demand is likely to lead to a far more comprehensive and denser distribution of data centre-like infrastructure than has been seen to date.

Valuation

Data centres are a unique type of commercial real estate. A key question to any investments, whether in the form of debt, equity, or acquisition, is what is the valuation? The question of data centre valuation is central to the investment calculation. The most pertinent consideration to the value of a data centre business is the reliability of the long-term revenue stream created from customer contracts and the utilisation of the facilities to maximise operational cost efficiencies. A data centre with long-term customer contracts and full occupancy will generate predictable revenue. Private financial investors tend to focus on data centres that provide facilities to those large hyper-scalers such as Microsoft, Google, Facebook, Amazon, Oracle, and other cloud providers. The credit risk[16] for these types of customers is considered low. These investment-grade facilities are typically state-of-the-art and purpose-built to meet the specific requirements of individual hyper-scalers, which may differ according to individual needs.

However, investors in data centres may need to consider a unique range of issues. At the same time, factors such as facility type, age, location, tenant covenant, and regulatory issues are undoubtedly important. In data centre valuation, other criteria such as power supply access, fibre-optic network connectivity and building cooling systems' reliability, which typically play only a marginal role in ascertaining the value of other property types, carry substantial weight.

Data centres are increasingly deemed critical infrastructure, with many jurisdictions seeking data sovereignty. The purchase of data centres by foreign entities tends to be among the most sensitive acquisitions, with new rules on data sovereignty increasingly applied. Considerations may include the need to make voluntary or compulsory notification to, approval from, or some other form of vetting by, the screening authority. There also may be constraints concerning the obligations of ownership requirements or other conditions concerning governance and data access and storage. Foreign investment controls may also impact which assets are available for investment by private capital, particularly sovereign wealth

---

16 Standard & Poor Global Ratings generally range from credit rating of AA+ to AA for these company types.

funds and other funds deemed to be government-owned or controlled, and the structures used.[17]

Data centres are still frequently viewed as a niche real estate asset class because the sector is relatively small and lacks maturity. Still, it also means that investors have focused on development and partnerships with operators to build new centres, giving rise to short-term supply risks. There is too little understanding of this market to allow informed pricing of the risks and returns. Access to independent market intelligence on sector data and the sourcing of transactions remains challenging. As the sector grows, transparency and liquidity are improving.

The Royal Institute of Chartered Surveyors (RICS) maintains[18] guidance on the critical factors in a data centre valuation assessment. Like all property, the value basis will depend on the valuation's purpose. In any data centre acquisition, a threshold real estate concern is whether the underlying land on which the digital infrastructure rests is owned or leased under terms. Where data centre land is owned, investors should verify that the target company has a clear title to the ground by conducting title of land searches. Investors should also review the target company's purchase agreement when purchasing the land. Beyond the fundamental ownership issues, investors must ensure that the land is appropriately zoned for data centre operation. The latter may be significant in FLAP-D[19] and Asian markets such as Singapore and Hong Kong, where land is generally scarce, and large land parcels suitable for data centre development and operation are even more limited.

Where data centre land is leased, potential acquirers need first to conduct the title of the land searches to ensure the lessor is the legal owner of the land. And second, to review the lease between lessor and lessee to understand its critical terms of time, change of control, landlord's consent rights, and step-in rights to understand how these terms could potentially impact the valuation of the acquisition. Investors should ensure that the proposed transaction will trigger no termination right. The deal will be jeopardised if the landlord can re-enter the premises and take back the land due to the proposed transaction. Suppose the lease gives the landlord consent or notification right as a result of the proposed transaction. In that case, investors could consider adding a closing condition or a closing deliverable in the purchase agreement that obliges the sellers to deliver such consent

---

17 Whether direct investment or through a managed fund.

18 Rics.org. (2023). Available at: https://www.rics.org/profession-standards/rics-standards-and-guidance/sector-standards/valuation-standards/valuation-of-data-centres [Accessed 1 May 2023].

19 FLAP-D is a short notation for Frankfurt, London, Amsterdam, Paris, and Dublin.

or notification prior to closing, with no material changes or demands to change the current lease terms.

It is for the investors to understand whether the lease permits the data centre operator to sublease the space to its customers, as customer contracts may be drafted in a manner which creates a sublease. If there is such a limitation, investors should seek to understand whether a structuring alternative, licensing the space to customers, is permissible under the lease. Investors ought to also be aware of the landlord's identity, whether the landlord is a government or private entity or a person. If the landlord is a government entity, policy considerations could impact the lease terms rather than purely economic factors. These could include preferential lease economics to encourage data centre development. Still, these may be coupled with draconian restrictions on the alienation of the lease, which may provide government landlords with the ability to renegotiate the lease or take back the land if transactions are not correctly structured.

It is because the real estate on which any data centre sits is a crucial component of the value of the data centre business; environmental due diligence ought to be carried out to the same degree as in any real estate deal.

Investors need to understand what environmental risks might exist on the property based on past uses and therefore ensure that liability is appropriately apportioned between lessor and lessee and factor the environmental risks into the valuation of the deal.

Assuming that the valuation is required for the purchase or sale of an interest in the property by an occupier or investor or for company accounts purposes or loan security, the basis should be market value. The market value of the data centre will be a factor of its ability to produce an income by way of rent, and fees, from occupiers, using the accommodation principally for data storage/transaction purposes. If the data centre is already let, the valuer will therefore take account of current and the expected future net income when assessing its value. The value of a vacant or owner-occupied building will be based on its potential to produce income.

The income approach to value is one of the traditional approaches to valuing data centre real estate. It takes a selected income stream and capitalises it into value. The direct capitalisation technique under the income approach requires calculating a single year's annual stabilised NOI and then applying a market-supported capitalisation rate to arrive at a value conclusion. This works well when a regular and relatively constant annual income stream is likely to remain stabilised over the investment period. It is relatively simple and reliable if sufficient market data is available.

A more complex technique is yield capitalisation, also known as discounted cash flow (DCF) analysis, which uses multiple years of NOI, discounted to a present value, and then adds in the present value of the future sale of the subject property at the end of the hold period. This works well with new properties and is still in lease-up mode or being upgraded or repositioned in the marketplace. The most commonly used valuation by actual buyers of data centres in their underwriting process.

There are other income-based techniques like gross income multipliers (GIMs) or even net income multipliers (NIMs), but these tend not to be used or given much weight by buyers in determining how much to pay for their acquisitions or even by assessors. Irrespective of which income-based valuation technique you select, you must first start by identifying the area of the power shell of the real estate and the appropriate rental rates to be used. The key to accurate data centre assessment is to focus on the powered shell's proforma market-based income stream[20] alone.

The evaluation of the rental value will depend upon the amount of equipment, usually the number and the capacity of racks, that the demised space may accommodate. The ability of the area to house equipment will be constrained by the:

a) amount of power available for the tenant's equipment
b) cooling capacity and power consumption of air-conditioning equipment in the space
c) amount of emergency power generation available to provide adequate backup to the required standard (usually on an $N + 1$ basis)
d) the bandwidth of data cabling

Another key point for investors to understand is that the real estate component of a data centre can often represent only 15–20% of the overall investment, with mechanical and electrical (M&E) infrastructure accounting for 80–85%. As a result, investors should ascertain how much 'useful life' M&E can provide before re-investment is needed to make an informed decision. Therefore, it is essential to understand the specification and conditions of the M&E systems and the responsibility the tenant has for the M&E equipment on an ongoing basis and at the end of the lease.

The other most crucial contributor to value is the amount of available power to the centre because of its direct relationship with its equipment capacity. The language used to define the cost in the data centre market is often expressed in dollars ($) or euros (€) per MW, whether it is construction cost or lease procurement.

Therefore, those technical considerations are likely to outweigh the physical dimensions of the demised space in an assessment of the ability of the area to generate rental income. The other primary factor will be income security, such as

---

20 The shell and core with power, but excluding tenant fit out.

the strength and mix of tenant covenants and the terms and length of leases. The evaluation of rental value will also consider:

a) Revenue generated from wayleaves and licences allowing occupiers to run cabling through the building outside their demised premises.
b) Profits on the provision of power and other services.
c) Letting of meeting rooms and ancillary office space; and
d) additional licences, such as income from roof-mounted satellite dishes.

All data centre valuations may consider geopolitical risk. Many jurisdictions worldwide have introduced legal constraints following Europe's path in introducing strict privacy restrictions. In certain regions,[21] data infrastructure is identified as a sector where transactions could potentially raise national security concerns. Investors need to understand the existing regulatory landscape and also probe what is on the horizon in terms of regulatory change which could impact their investment. Understanding the risks of non-compliance to avoid poor uptake by risk-reluctant customers, liability under customer contracts or very significant fines will prevent or mitigate expensive mistakes.

From data centre customers' perspective, in any sale or merger transaction, it is crucial to make sure that their data is continued to be stored securely. Therefore, data centre customers may request considerable detail about data centre potential new operators' security policies and protections and may request that some specific protections be included in their contracts. From the investors' point of view, ensuring customer contracts foresee liability caps would be prudent to avoid uncertainty.

One of the other reasons that colocation, particularly wholesale colocation, has become so popular is the significantly lower upfront capital investment and predictable and transparent costs for initial commitments, monthly expenses, and future expansion options. Moreover, the time-to-market cost for additional capacity or new locations is much lower than if the organisations were to try to acquire land, submit building plans for approval, and construct their own facility. Because IT strategy can change quickly, companies value fluidity with their data centre infrastructure. A company's data centre may fit its needs today but could be inefficient later. Colocating provides flexibility and helps users avoid getting stuck in a solution that does not fit their needs.

Companies also save on outsourcing specialised skill sets. Data centre operation requires a level of expertise that many companies often lack. While it is possible for companies to develop staff to fill these roles, outsourcing the requirement is

---

21 GOV.UK (n.d.). *National security and investment act 2021.* [online] Available at: https://www.gov.uk/government/collections/national-security-and-investment-act.

often faster, less expensive, and more efficient. These data centre providers are colocation experts who may provide specialised solutions that best fit their customers' needs.

There are also economies of scale, and data centre providers are experts in designing and building data centres and often do it cost-effectively. Large providers can also leverage their size to lower construction and power costs, which may be passed on to the user, creating lower operating expenses than owning the data centre. Data centre providers offer various services to meet their users' needs and ease of customisation. They can also use their scale to attract third-party service providers, which creates a valuable ecosystem that is hard for single users to replicate. A colocation data centre may often have a more substantial fibre infrastructure and easier access to cloud service providers, giving users low latency to their cloud environments and the end customer.

Growing your data centre presence is easier with a colocation data centre provider. It also allows local businesses to expand to a global footprint agilely. Many colocation data centre providers have global footprints; therefore, a customer of these colocation providers may easily penetrate new markets quickly. The relationship between a user and a data centre provider is typically seen as a long-term partnership. Should a company need a data centre in a new market, they can often deploy infrastructure in their provider's facility in that region. Providers like Digital Realty, Equinix, and CyrusOne report the vast majority of their customers to have deployments in more than one of their data centres and many in more than one country. Even large tech companies like Microsoft, Facebook, and Amazon find leasing data centre space a viable strategy in addition to owning and operating their own infrastructure.

When considering colocation providers, the two primary qualifiers are the amount of power and space required. While every colocation provider has a different threshold, they may differ depending on their available space inventory and remaining power capacity for a given facility or market. In addition, it will matter if the suite, pod, hall, or facility you need is in one of the first of many buildings yet to be built on a new campus or if it is already fully tenanted.

Financial due diligence is critical; while some colocation providers are well-funded, others may be highly leveraged. When considering a data centre provider, whether wholesale or retail, consider the size, financial stability, and operational experience of the organisation. Moreover, site visits and speaking with existing customers are recommended to get operating history and response to any issues. Unlike software, physical data centres cannot simply be upgraded with a code patch, hot fix, or next release.

Typically, investors are interested in colocation-type data centres. This market is split primarily between retail colocation and hyper-scale wholesale colocation as outlined earlier.

Colocation Leases

The types of colocation leases fall into various categories driven by the size, capital availability, and user requirements of facility operations. Often the user chooses between a company-owned data centre and a leased space. These considerations often involve the analysis of the cost of ownership and capital.

Colocation leases can range from several servers to an entire data centre. Depending on user needs and lease size, data centre providers may prefer to structure leases differently.

Smaller footprints, leases of 50 kW and less are usually all-in leases,[22] where the user pays a set price per month with slight variation. These leases may be for several racks or spaces within frames, and the type of lease may depend on the power needs and the desire and experience of the customer in maintaining their own IT gear. The lease price includes both the rental rate and power cost.

Leases of between 50 kW and 5 megawatts (MWs) are often modified gross leases; typically, these are real estate leases that include at least some pro rata share of the cost of the OPEX in the base rent. Typically, the pro rata variable is the cost of electrical power, where the user pays a set price per kW of data centre infrastructure they lease per month, plus the cost of the energy they use.

A triple net (NNN) lease is generally for a more significant lease of 5 MW and higher and applies where the real estate lease passes through all of the customer's share of the operating expenses, both shared and unshared. These tend to be for the more sophisticated user and may involve leasing an entire building or an autonomous suite. Generally, the tenant procures all fibre connections, utilities, and maintenance services. The tenant or landlord may own part or all of the facility infrastructure, but the tenant is obliged to maintain the facility. If the tenant owns the infrastructure, the landlord may require the tenant to adhere to a prescriptive maintenance schedule. These leases tend to be similar to industrial leases and are generally long-term, often 10–15 years in duration, with options to renew and purchase, with strict obligations on surrender. It is not uncommon for these to be built by the landlord for a pre-let tenant. The base rent may be based on the square footage/meterage of the rental area or the power availability.

Another variation on a triple net lease is to have a powered core and shell with a lease of the building or autonomous suite. The landlord may provide the raised floor/fibre connectivity/power supply. Whether the tenant or the landlord owns the facility infrastructure, the tenant may be obliged to provide the maintenance. The tenant installs and maintains all the power and networking distribution, racks, and IT gear. Fibre access may be direct to fibre providers or through an

---

22 Sometime termed 'Gross/Full-Service'.

MMR.[23] In this model, the landlord provides limited services and typically no SLAs. Base rent may be based on the square footage/meterage, but in this model, it is more likely to be based on power availability.

Wholesale Colocation

This type of colocation involves the leasing of the building or autonomous suite in which the landlord provides the fibre connectivity, conditioned power at either the demarcation at the PDU[24] or RPP[25] level, and environmental controls such as cooling and humidity.

The landlord usually owns and maintains facility infrastructure. In this model, the tenant usually installs and supports all power distribution downstream from PDU/RPP, networking distribution, racks, and IT gear. The fibre access may be direct to fibre providers or via the MMR. The environmental controls and facility maintenance are governed by SLAs. The base rent is usually based on power.

Retail Colocation

This type of colocation involves leasing a portion of the building or a suite, which may also be allocated as a caged space, rack, or space within a single frame. This space may be secured via a lease or a licence. It is typical for the landlord to provide fibre connectivity; it may provide networking and Internet access, conditioned power at server-usable levels to supply the equipment in the racks, environmental controls, cage security, and usually the rack itself with all the associated maintenance. It is then for the tenant to install and maintain its own IT gear. SLAs are customarily applied to assure environmental controls; they may also cover connectivity. The base rent may be gross, modified gross, or NNN. However, a NNN lease is rare. Power charges may be based on power capacity, whether or not used, or actual usage.

---

23 A 'meet-me room' (MMR) is typically found with the data centre, both colocation and owner operated. It provides a managed and secure space for interconnection of telecommunications companies' carrier services to the facility and owner's internal network without incurring local loop fees.

24 A PDU is used to distribute power, typically to networking and computer hardware housed in a rack in a data centre. A basic PDU has one input and multiple outputs, each designed to supply power to one piece of equipment. A stable power supply is critical in data centres.

25 A remote power panel (RPP) provides power distribution extensions from PDUs or other power sources directly to server racks.

A triple net lease may absolve the owner of the most risk, and the tenant may pay even the costs of structural maintenance and repairs to supplement the rent, property taxes, and insurance premiums. When these additional expenses are passed on to the tenant, the property owner generally charges a lower base rent. There are quite a few variants of this type of lease. In a single net lease, the property owner transfers a minimal risk to the tenant, who compensates the property taxes, often described as a net lease or an 'N' lease. Double net leases, which are also known as net-net leases or 'NN' leases, are notably widespread in commercial real estate. For these types of leases, the tenant pays property taxes and insurance premiums in supplement to the rent.

Service-Level Agreements (SLAs)

The purpose of an SLA is to provide the tenant with a remedy for failure to maintain agreed-upon service levels. It is often preferred that a failure to meet a service level should not be a lease breach but a compensatory event. Typical SLA covers power, internal environment, notices, staffing, and security. The SLA on power will usually cover power availability and outage scenarios, the number of disruptions, the length of each disruption and monitoring. The environmental SLAs cover the temperature range, humidity range, and sensor's location with specified monitoring protocol. The SLA will also prescribe the notice time for incidents, the availability and response time following an incident, and the time to respond and identify the root cause and suggest a solution. It may also deal with many security breaches. SLA remedies may include rent credits or abatement, often tied to a percentage of monthly rent that may be capped. Termination rights are typically included for chronic repeat failures but exclude force majeure or planned maintenance tenant work or changes.

Managed Hosting and Cloud Services

First, these arrangements are not leases. In managed hosting, the provider owns and maintains the IT gear whilst the user provides and updates software. Cloud/SaaS[26] may provide and update software. Managed hosting is also described as managed dedicated hosting, dedicated servers, or single-tenant hosting. Those names all refer to an IT service model where a customer leases hardware like the

---

26 SaaS Hosting is a type of Web hosting service wherein your software application and files are stored on a hosting provider's servers.

servers, rack storage, and network from a managed service provider (MSP). In this model, the hardware is dedicated to just that customer, consequently the 'single tenant' moniker, giving access to the full performance capabilities of the hardware. Customers have total control over their hardware, applications, operating systems, and security as a single tenant. The 'managed' part means the MSP handles the administration, management, and support of the infrastructure, which is housed in a data centre owned by the provider instead of onsite at a customer's location. Some companies prefer to lease data-centre space and hardware from a provider and manage it themselves, referred to as hosting or unmanaged hosting. However, as technology continues its rapid innovation, many organisations find that outsourcing day-to-day infrastructure and hardware vendor management to an MSP is of higher value to their business than managing it themselves.

Cloud services negate the initial need to purchase and operate hardware and offer lower upfront costs. However, while the pricing model for cloud services can appear relatively simple at first with 'pay for use service costs', it can be somewhat misleading. If the pricing model implications are not fully understood, it can cause some organisations to underestimate the fixed recurring and usage-based costs. The primary charges may be complex and often fall into four general categories: compute, data storage, data read-and-write to and from data storage, and upload and download bandwidth. Depending on the nature of the application, the size of the stored data and the amount and type of traffic over time, this can become substantially more expensive than originally projected. While cloud services may be more cost-effective initially, at some point, as computing demands increase in magnitude, the relatively fixed recurring model of the wholesale data centre may become lower than cloud-based services.

Total Cost of Ownership (TCO)

While TCO is essential in any decision, and the cost of colocation and cloud services may appear to diverge as market conditions evolve, initial price alone should not be the only factor. Wholesale colocation provides secure control of your facility and IT resources, while the cloud offers on-demand capacity, but most often, is also a shared resource platform with less assurance of direct control and with security concerns that may not be acceptable for some organisations. The cost of wholesale colocation is fixed, and the customers are free to maximise their IT performance capacity based on the investment of their own IT equipment, rather than paying per gigabyte for recurring cloud service usage-based costs for computing, and data storage, data read-and-write, and upload and download bandwidth.

There are also management benefits. In addition to the lower costs, dealing with a large-scale provider avoids multiple interfaces with management. These contractual and logistical problems commonly occur when dealing with numerous colocation providers. This is especially true when dealing internationally with different providers in each country. The added value of centrally managing all contracts and the ease of locating additional space in any facility in a single provider's global network of data centres should not be overlooked. Moreover, the ability to continuously measure and track conformance with SLAs via transparent access to consolidated, centralised performance, and status monitoring helps ensure technical and financial compliance.

The decision to build is much more demanding. Besides the purchase of large tracts of land, which may involve hundreds of acres, wholesale providers will have proposed a master plan to be able to scale up over time when reserving and committing to, and in some cases, competing for commitments for essential resources such as power and water. Other than top-level Internet hyper-scale giants, even well-funded and established enterprise organisations may not be able to compete based only on their own data centre requirements, especially if it is only in the range of 2–5 MW. They most likely will not have the available financial resources or choose not to apply them to match the purchasing power of a significant wholesale colocation provider. Moreover, they may also lack the breadth and depth of experience and the scale of internal resources to coordinate and support the building process from beginning to end.

Merger and Acquisition

Data centre merger and acquisition predictably involves a considerable amount of due diligence by the buyer. Before executing any transaction, the buyer will want to make certain that it knows what it is buying, what obligations it accepts, the extent and nature of the seller's provisional liabilities, problematic contracts, intellectual property issues, litigation risks, and much more. Data centre's due diligence in the colocation and service provider sector requires more than just evaluating power and cooling infrastructure. After signing non-disclosures and confidentiality agreements, consulting and engineering services companies are asked to visit some or all of the data centre assets being considered acquisitions. Alternatively, due to the urgency and pace of pending M&A deals, undertaking a desktop document evaluation of the relevant drawings and documentation they have obtained applicable to the site(s) is often preferable.

Critical facility's due diligence endeavours and opinions usually include evaluations and assessments of existing asset conditions, equipment age, building management systems, CAPEX and OPEX, tier levels/equivalents, energy efficiency, and any possible points of failure. It is the case that a rough order of magnitude (ROM) cost estimate for expansion, upgrades, or improvements identified in the course of analysis is often required to assess the various 'what if' return on investment modelling.

A due diligence endeavour should include additional issues beyond the physical asset. These include operational procedures, maintenance management, and staffing. It is critical to review the historical failure incident report logs to identify reasoning for past failures that typically may result from human error or, in some cases, may identify a more critical defect issue with the building infrastructure and thus influence the valuation. These factors drive costs and are frequently important factors in data centre outages that impart reputational damage.

It is often critical to review the IT systems architecture to identify the applications, security, and processes for both the acquirer and the client-facing customer. These features are key revenue drivers and essential to the SLAs. The likely synergies, integration costs for multiple systems, and risks to the overall data centre should be considered. Understanding these critical aspects of a deal during the due diligence phase is vital.

A systems architecture assessment is essential due to the tight integration and effect that it may have on the operations, management, and the end user client, whether it is the external or internal facing environment. The enterprise and operations backend may seriously impact services' underlying effectiveness and total ownership cost.

The choices of the operations management suites,[27] monitoring tools, client portals, and use of cloud, SaaS,[28] and PaaS[29] all have definitive impacts on CAPEX and OPEX and the reliability and resiliency of the facility asset. This interrogation goes to the value of the purchase or uncovers the potential future investments you may need. Without this additional analysis, the acquirer may not have as accurate an understanding of the value of an asset as they should make informed decisions.

As data centre investment volume continues to gain impetus and the sector advances into a mainstream real estate asset class, demand for accurate and

---

27 Odika, C. (2018). *Microsoft Operations Management Suite Cookbook: Enhance Your Management Experience and Capabilities Across Your Cloud and On-premises Environments with Microsoft OMS*. Birmingham: Packt Publishing Limited.

28 Software-as-a-Service (SaaS) applications run in the cloud. Users subscribe to SaaS applications instead of purchasing them, and they access them over the Internet.

29 Platform as a service (PaaS) is a complete development and deployment environment in the cloud, with resources that enable you to deliver everything from simple cloud-based apps to sophisticated, cloud-enabled enterprise applications. You purchase the resources you need from a cloud service provider on a pay-as-you-go basis and access them over a secure Internet connection.

reliable evaluations is mounting in what is widely thought[30] to be one of the most complex fields of real estate valuation.

Factors such as facility type, age, location, tenant covenant, and regulatory issues are undoubtedly important in data centre valuation. Other criteria, such as power supply access, fibre-optic network connectivity, and reliability of building cooling systems, most of which typically play only a marginal role in ascertaining the value of other property types, carry substantial weight in data centre valuation.

The valuation of a data centre will need a significant element of judgement in arriving at the valuation figure, backed by comprehensive knowledge and experience of this type of property and a capacity to analyse the complex cash flows that data centres frequently produce.

Those investors who understand these key issues related to data centre investments will be best positioned to benefit from a competitive digital infrastructure landscape in the coming years.

While the prospect of investing in data centres might seem formidable, this sector is also abundant in opportunities for investors willing to familiarise themselves with this alternative asset class.Driven by the growth of cloud computing, IoT and Big Data, the demand for data centres is at an unprecedented high and will only keep growing. As smart cities develop further, they demand the backbone of a highly functional and heavily invested data centre platform to be successful.

This growing demand, coupled with the high level of expertise needed to operate data centre assets, presents numerous opportunities from an investment perspective. For instance, a successful new operator is highly likely to be acquired by a large global data-centre operator in the medium term. Even if the operator is unsuccessful but has a facility that matches critical criteria, it is still very likely that other data centre operators may compete to take over the business.

It is most likely that investment opportunities in the data-centre sector will abound as Industry 4.0 and IoT march relentlessly onward, but to be successful, investors must develop a high level of comfort with the asset class. Investing in this sector requires flexibility, in-depth knowledge, and a global perspective.

---

30 Cbre.com.au (2021). *Valued insights: Sustainability and data centre valuation.* [online] Available at: https://www.cbre.com.au/insights/articles/Valued-Insights-Sustainability-and-Data-Centre-Valuation [Accessed 1 May 2023].

9

Sustainability

Corporate Sustainability

The United Nations (UN) defines sustainable development as 'development which meets the needs of the present without compromising the ability of future generations to meet their own needs.'[1] It recognises that through their activities and business relationships, organisations can affect the economy, environment, and people and, in turn, make negative or positive contributions to sustainable development. In 2015, all UN Member States adopted the 2030 Agenda for Sustainable Development,[2] which provides a shared blueprint for peace and prosperity for people and the planet, now and into the future. At its heart are 17 Sustainable Development Goals (SDGs)[3] shown in Figure 9.1, which are an urgent call for action by all countries – developed and developing – in a global partnership. They recognise that ending poverty and other deprivations must go hand-in-hand with strategies that improve health and education, reduce inequality, and spur economic growth – all while tackling climate change and working to preserve our oceans and forests.

Organisations are facing increasing pressure from governments, investors, their customers, and the public to operate in a more environmentally and socially

---

1 United Nations Brundtland Commission (1987). *Report of the world commission on environment and development: Our common future towards sustainable development 2. Part II. Common challenges population and human resources 4.* [online] Available at: http://www.un-documents.net/our-common-future.pdf.

2 United Nations (2015). *Transforming our world: The 2030 agenda for sustainable development | Department of Economic and Social Affairs.* [online] United Nations. Available at: https://sdgs.un.org/2030agenda.

3 United Nations (2015). *Sustainable Development Goals.* [online] United Nations Sustainable Development. Available at: https://www.un.org/sustainabledevelopment/sustainable-development-goals/.

Data Centre Essentials: Design, Construction, and Operation of Data Centres for the Non-expert, First Edition. Vincent Fogarty and Sophia Flucker.

Figure 9.1 UN Sustainable Development Goals. *Source:* Reproduced with permission of UNITED NATIONS.

responsible manner; it has become a business imperative.[4] Investors may screen companies based on their environment, social, and governance (ESG) performance. Greenpeace's Clicking Clean campaign highlighted the renewable energy content of major digital platforms.[5]

It has become commonplace for organisations to share their green credentials. However, this is often greenwashing rather than transparent and verifiable reporting. The Global Reporting Initiative (GRI)[6] aims to provide transparency on how an organisation contributes/aims to contribute to sustainable development and can form a framework for defining a company's sustainability strategy through the use of standards. These are aligned with the SDGs. Other sustainability reporting initiatives are available such as UN Global Compact,[7] World Business Council for Sustainable Development (WBCSD), and Greenhouse Gas Protocol.[8]

4 Pole, S. (2019). *Q&A: A Practical Guide to Science-Based Targets.* [online] Sustainable Brands Available at: https://sustainablebrands.com/read/business-case/q-a-a-practical-guide-to-science-based-targets#:~:text=Convincing business case. [Accessed 8 May 2023]
5 Greenpeace International (2019). *Greenpeace international.* [online] Available at: https://www.greenpeace.org/international/publication/6826/clicking-clean-2017/.
6 GRI (2019). *GRI Standards download homepage.* Available at: https://www.globalreporting.org/standards/.
7 United Nations (2019). *Homepage | UN Global Compact.* Available at: https://www.unglobalcompact.org/.
8 World Business Council For Sustainable Development and World Resources Institute (2004). The Greenhouse Gas Protocol: A Corporate Accounting and Reporting Standard. Geneva, Switzerland; Washington, DC: World Business Council For Sustainable Development.

In the EU, the Corporate Sustainability Reporting Directive (CSRD) entered into force on 5th January 2023. This new directive modernises and strengthens the rules concerning the social and environmental information that companies have to report. A broader set of large companies, as well as listed SMEs, will now be required to report on sustainability – approximately 50 000 companies in total. An Assessment Framework for Data Centres, applied in the context of European Taxonomy Regulation will form the basis of mandatory reporting under the CSRD (this is derived from the EU Code of Conduct for data centre energy efficiency).[9,10,11]

Data centres have a high and growing environmental impact due to their high energy consumption and use of resources, and hence discussions around data centre sustainability tend to focus on this area. Organisations commonly hold ISO 14001 certification.[12] ISO 14064-1[13] specifies the quantification and reporting of greenhouse gas emissions and removal. When analysing emissions for accounting and reporting, these are commonly categorised into three scopes:

Scope 1 emissions	Direct emissions from sources owned or controlled by the company, e.g. combustion of fuel in boilers or vehicles, fugitive emissions from refrigeration equipment
Scope 2 emissions	Indirect emissions from the generation of purchased electricity, steam, heating, and cooling in activities owned/controlled by the company
Scope 3 emissions	All indirect emissions relating to upstream and downstream activities as a consequence of activities of the company but not owned or controlled by the company, e.g. use of sold products/services

Organisations may set commitments and targets, e.g. SBTi,[14] Climate Neutral Data Center pact[15] which includes commitments for energy efficiency, clean energy, water, circular economy, and circular energy systems. The Irish government has stated that 'it is expected that data centres will align with the EU Climate

---

9 eur-lex.europa.eu. (n.d.). EUR-Lex - 32022L2464 - EN - EUR-Lex. [online] Available at: https://eur-lex.europa.eu/eli/dir/2022/2464/oj.

10 Bertoldi P., Assessment Framework for Data Centres in the Context of Activity 8.1 in the Taxonomy Climate Delegated Act , European Commission, Ispra, 2023, JRC131733.

11 eur-lex.europa.eu. (n.d.). EUR-Lex - 32020R0852 - EN - EUR-Lex. [online] Available at: https://eur-lex.europa.eu/eli/reg/2020/852/oj.

12 ISO 14001:2015 Environmental management systems – Requirements with guidance for use.

13 ISO 14064-1:2018 Greenhouse gases – Part 1: Specification with guidance at the organisation level for quantification and reporting of greenhouse gas emissions and removals.

14 World Resources Institute (n.d.). *The Science Based Targets initiative (SBTi).* [online] Available at: https://www.wri.org/initiatives/science-based-targets.

15 *Climate neutral data centre pact – the green deal need green infrastructure.* [online] Available at: https://www.climateneutraldatacentre.net/.

Neutral Data Centre Pact energy efficiency and water use targets and set themselves targets to achieve zero-carbon electricity use at all hours'.[16]

Energy Consumption and Energy Efficiency

Data centres are large energy consumers, accounting for an estimated 1% of global electricity consumption.[17] In Ireland, data centres were recorded as consuming 11% of total metered electricity consumption in 2020[18] and have been forecast to account for 29% of all electricity demands by 2028.[19] In most cases, the majority of energy consumption is used on computing, i.e. powering the servers and other IT hardware housed in the facility. Although new generations of hardware are more efficient than previous ones in terms of computation per watt,[20] this does not equate to the reduced energy consumption of IT hardware – demand continues to outstrip efficiency gains. The remaining power is used in cooling the IT equipment and in power distribution systems, e.g. UPS losses.

In recent years, the industry has become more aware of the opportunities to improve energy efficiency and reduce energy consumption. As energy prices increase (for example, due to recent geopolitical events in Europe), the business case strengthens – saving energy saves operating costs and increases profitability. There is a perception, however, that saving energy is not compatible with reliability. The requirement for high availability is often used as an excuse for poor energy performance, for example, because there is a perceived risk of operating at higher temperatures. In fact, there are many ways in which to operate redundant systems in a manner which is efficient and does not increase risk. One example of this is running all cooling unit fans at a reduced fan speed, rather than running some at full speed with others switched off. The energy consumption is less (due to the cube law) and there is less wear on components operating at lower speeds. In the event of unit failure, the remaining units are already

16 www.gov.ie (n.d.). *Government statement on the role of data centres in Ireland's enterprise strategy.* [online] Available at: https://www.gov.ie/en/publication/5e281-government-statement-on-the-role-of-data-centres-in-irelands-enterprise-strategy/ [Accessed 4 Oct. 2022].

17 Masanet, E., Shehabi, A., Lei, N., Smith, S. and Koomey, J. (2020). Recalibrating global data center energy-use estimates. *Science*, 367 (6481), pp.984–986.

18 www.cso.ie (n.d.). *Key findings – CSO – Central Statistics Office.* [online] Available at: https://www.cso.ie/en/releasesandpublications/ep/p-dcmec/datacentresmeteredelectricitycons umption2020/keyfindings/ [Accessed 5 Mar. 2023].

19 *All-Island Generation Capacity Statement 2019–2028,* SONI Ltd, EirGrid Plc 2019.

20 Koomey, J.,Berard, S., Sanchez, M. & Wong, H. (2011). Implications of Historical Trends in the Electrical Efficiency of Computing. *Annals of the History of Computing, IEEE*, 33, 46–54. 10.1109/MAHC.2010.28.

running and just need to increase their speed.[21] Often, the designer has focussed on sizing equipment for full load operation and adjustments are required to optimise operation at part loads (likely to be most of the operating life of the facility).

Best practices for improving data centre efficiency are well-documented, for example, in the European Code of Conduct for Data Centre Energy Efficiency Best Practices.[22] Most facilities apply a number of these practices in design and operation, although there continues to be room for improvement in many cases. As is commonly observed in a variety of building types, data centres also experience a performance gap between how they are designed to operate and how they do in reality.[23]

There are many areas in which energy efficiency can be improved. Typically, the highest priority/largest opportunity is the IT equipment energy consumption. This includes several facets which relate to right-sizing and re-architecting[24]:

- Software efficiency.[25] The Sustainable Digital Infrastructure Alliance has a steering group working on creating the tools and frameworks to display the environmental footprint of applications to end users enabling them to take action and reduce the footprint.[26]
- Virtualisation – i.e. distributing hardware resources across several processes, rather than having dedicated hardware for each.
- Rationalisation and consolidation – i.e. managing applications and assets to avoid zombie servers (powered on with little or no productive activity[27]).
- Hardware efficiency, including that of power supplies.[28]

Addressing this can have a significant impact on energy consumption and hardware footprint. A reduction in IT power consumption will have a knock-on effect on power and cooling energy consumption. The industry commonly uses

---

21 Flucker, S. and Tozer, R. (2012). *Data Centre Energy Efficiency Analysis to Minimise Total Cost of Ownership*. BSERT

22 Acton, M., Bertoldi, P., Booth, J. (eds.), (2023). *Best Practice Guidelines for the EU Code of Conduct on Data Centre Energy Efficiency*. European Commission, Ispra, 2023, JRC132576.

23 CIBSE (2020). *TM61: Operational performance of buildings*.

24 Sundberg, N. (2022). *Sustainable IT Playbook for Technology Leaders*. Packt Publishing.

25 Greensoftware.foundation (2021). *Green Software Foundation*. [online] Available at: https://greensoftware.foundation/.

26 sdialliance.org (n.d.). *Defining the environmental footprint framework for server-side applications*. [online] Available at: https://sdialliance.org/steering-groups/assessing-the-digital-carbon-footprint.

27 The Green Grid (2017). *WP74 solving the costly zombie server problem*.

28 www.energystar.gov (n.d.). *Enterprise servers*. [online] Available at: https://www.energystar.gov/products/enterprise_servers [Accessed 5 Mar. 2023].

the metric power usage effectiveness (PUE)[29] to describe how much of the total energy consumption of the facility is used by the IT equipment and how much is required for cooling and power distribution. PUE is a ratio:

$$PUE = \frac{Total\ Energy\ Consumption}{IT\ Energy\ Consumption}$$

Total energy consumption includes IT energy consumption plus the energy consumed by cooling systems and power distribution losses. Therefore, the theoretically ideal value = 1, where all energy is used for IT only and power and cooling losses equal zero. The closer to 1, the better. Note, this metric provides no information on how efficiently the IT energy is being used, only how much energy is being used in cooling and power distribution. The standardised definition defines this as energy consumption over a period of one year. However, it can be useful to undertake an analysis of instantaneous values to understand how this varies with IT load, operating mode, and ambient temperature. A PUE = 2 would mean that as much energy is used in power and cooling as reaches the IT equipment. This would be typical for a legacy facility where the power and cooling systems are not optimised for energy performance. A low-energy efficiency design may target a PUE of around 1.2 or less. Figure 9.2 indicates the breakdown of energy for each of these cases:

The main difference between the two cases is the reduction in the mechanical system's energy consumption, particularly the compressors (used in refrigeration) and cooling unit fans. This reduction is achieved by operating at higher temperatures, increasing the operational time spent in free-cooling (economiser) mode and reducing cooling unit fan speeds. It is easier to achieve low PUEs when the systems have been designed to be low-energy. However, significant PUE improvements are still possible on live, legacy facilities, particularly when operating at part load (see Chapter 7). Of course, any changes must be managed in a manner that mitigates risk. For air-cooled facilities, air management is an important enabler to facilitate energy savings. See Chapter 5.

PUE is often used as a marketing tool – companies may boast about their design rather than achieved value and claim that their facility is sustainable even though this only describes one aspect of their energy performance.

Colo pricing may not be transparent and the overhead cost of power and cooling systems may be passed on to the customer with a mark-up, which does not incentivise savings.

---

29 ISO/IEC 30134-2:2016 *Information technology – Data centres – Key performance indicators – Part 2: Power usage effectiveness (PUE).*

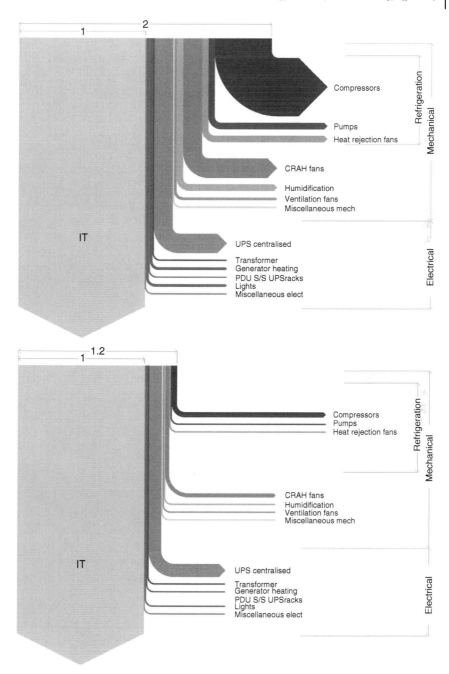

Figure 9.2 PUE breakdown.

IT power usage effectiveness (ITUE) and total power usage effectiveness (TUE) have been proposed as metrics which account for the power distribution and cooling losses inside IT equipment.[30] This is even more relevant for liquid-cooled IT hardware which does not include fans for air movement.

Renewable Energy

It is not just the amount of energy that is being consumed that is important but also how polluting that energy is. Electricity generated from fossil fuels has a much higher carbon footprint than that generated from renewables. Most data centres are connected to the regional or national electricity grid, so will be subject to that local grid carbon intensity. Two identical facilities in different countries could have very different environmental impacts due to their energy supplies. Many of the larger data centre companies advertise their commitment to purchasing renewable energy. Microsoft has committed to 100% supply of renewable energy by 2025, meaning they will have power purchase agreements for green energy contracted for 100% of carbon-emitting electricity consumed by all of their data centres, buildings, and campuses.[31] This is an important step in operating more sustainably but there are a couple of limitations with this:

1) In most locations, there is a limited amount of renewable energy. Increased consumption of renewable energy does not automatically increase capacity (additionality). Some data centre operators directly invest in specific renewable projects to generate green energy capacity matching their consumption.
2) If the data centre is connected to the grid, the actual electricity being consumed will be whatever the grid energy mix is.

On-site renewable generation is rare for several reasons:

1) Cost – capex and opex.
2) Skills and expertise. Design, operation, and maintenance of power generation require different expertise and add complexity.
3) Space limitations – generation plant takes up valuable real estate. Covering the roof with solar panels may only provide a fraction of the total energy

30 Patterson, M.K., Poole, S.W, Hsu, C-H., Maxwell, D., Tschudi, W., Coles, H., Martinez, D.J. & Bates, N. (2013). TUE, a new energy-efficiency metric applied at ORNL's jaguar. Supercomputing. ISC 2013. Lecture Notes in Computer Science, vol 7905. J.M. Kunkel, T. Ludwig and H.W. Meuer (Eds.). Berlin, Heidelberg: Springer. https://doi.org/10.1007/978-3-642-38750-0_28.
31 Microsoft (2021). *Environmental sustainability – microsoft CSR.* [online] Available at: https://www.microsoft.com/en-us/corporate-responsibility/sustainability? activetab=pivot_1%3aprimaryr3.

requirements depending on location, area and site density (also may require structural reinforcement for existing roofs and modifications for maintenance access).

Generators

On-site standby generators are used in nearly all data centres to provide backup power in the event of mains failures (see Chapter 5). Most burn diesel (a fossil fuel). In the United Kingdom, this includes a proportion of bio-diesel content. Some facilities use hydrotreated vegetable oil (HVO), which is derived from vegetable oil, used cooking oil, or other non-fossil fuel lipids. Testing suggests lower particulate matter and smoke emissions, comparable NO_x emissions, and slightly higher fuel consumption by volume but generally comparable performance.[32] The rate of adoption of HVO is currently limited by availability and higher cost. The use of generators for grid peak management increases emissions, particularly in cases where the grid is comprised of a high renewable content and has a low carbon intensity. Generator emissions are covered by regulations including the EU ETS.[33] In future, it may become commercially viable to use hydrogen as a fuel source which will help decarbonise energy generation.

Water Usage

Many data centres have been able to reduce the energy consumption of their cooling systems by using adiabatic, evaporative cooling, which uses the cooling effect of water evaporating rather than electricity-powered refrigeration. The impact of this is increased water consumption on site. Clearly, this is not a desirable effect in locations that suffer from water shortages/seasonal droughts. However, the net water usage may be less when the water used to generate electricity is considered. The net water usage may be more, but the overall environmental impact may still be less when considering the reduced environmental impact of using less electricity. A life-cycle approach is required in order to analyse the trade-offs.

---

32 Hawes, N. (2022). *Generator Set Performance on HVA Fuel: QSK95 Test Report Summary.* Cummins Inc. https://www.cummins.com/white-papers/generator-set-performance-hvo-fuel-qsk95-test-report-summary.
33 climate.ec.europa.eu (n.d.). *EU Emissions Trading System (EU ETS).* [online] Available at: https://climate.ec.europa.eu/eu-action/eu-emissions-trading-system-eu-ets_en.

Heat Recovery

Reusing waste heat is one way that data centres try to improve their environmental credentials. Rather than rejecting heat from cooling processes to the atmosphere, this can be captured and used by others (residential buildings, greenhouses, and swimming pools), for example, via a heat network, thereby reducing their energy consumption of heating. There are a few barriers to the wider adoption of waste heat reuse:

1) Although there may be a large amount of heat available, this is relatively low-grade heat, e.g. air <40 °C. This limits the economics, applications, and distance of the heat user. It is easier to look for an external user for waste heat during the facility site selection process and, of course, there is more need for heating in locations with colder climates. Note this is one of the selling points of liquid cooling – the ability to recover higher temperature heat from the data centre.

2) The data centre will be rejecting heat all year round but for many applications, heating is seasonal. A solution must be in place to deal with heat rejection 100% of the time. This may mean that both waste heat recovery and a traditional heat rejection system must be installed, creating additional infrastructure and cost.

3) Risk and contractual commitments. Heat recovery systems and plants are not standard in data centre installations; it is unlikely that the data centre operations team would have the skills and resources to maintain such an installation, meaning that a third party would need to be responsible. In many cases, heat pumps are required to increase the recovered heat temperature in line with the heat user's requirements. The heat user may also need to be tied in to certain commitments about how much heat they would consume and when. It is easier to manage these aspects in locations with existing district heating networks.

There are examples of heat recovery systems being installed as part of a planning requirement but then never being connected (waste of capex and embodied environmental impact). The SDIA argues that the business case for waste heat recovery needs to be developed by looking at the economics and selling this resource to energy companies who have the interest and means of using it.[34]

Heat recovery for on-site use is possible, although the heating requirements in the data centre building are typically only a small fraction of the heat rejection. In most cases, a standard heating solution will be used such as electric heating of data centre office areas.

---

34 Judge, P. (2022). *Heat Reuse: It's Time It Meant Business*. DatacenterDynamics. https://www.datacenterdynamics.com/en/analysis/heat-reuse-it-is-time-it-meant-business/.

Life Cycle Impacts

Energy consumption during the operational phase is only one part of a data centre's environmental impact.[35] Life cycle assessment (LCA) is a methodology which considers the holistic impact of a product, process, or service on the environment. An LCA looks at the products and processes within a system from cradle to grave, from the extraction of raw materials through manufacturing, transportation, operation, and eventual disposal, as indicated in Figure 9.3.

The principles, framework, requirements, and guidelines are set out in international standards BS EN ISO 14040[36] and 14044.[37] Currently, missing from most organisations' assessment of their data centre environmental impacts are the embodied impacts, which include the extraction and transformation of raw materials to fabricate the IT equipment, infrastructure (building services and network), the construction of the building itself and the disposal of these materials and equipment at end of life. The problem with this is that an important part of the data centre's environmental impact is ignored.

For example, printed circuit boards (PCBs) and integrated circuits (ICs) found in IT equipment contain numerous raw materials, many of which are finite, environmentally polluting to extract and treat at end of life (if sent to non-regulated countries), have poor to no infrastructure for reclamation at end of life, have high risk and politically volatile supply chains, and are socially impactful. Complex supply chains mean traceability of components is difficult; this lack of transparency

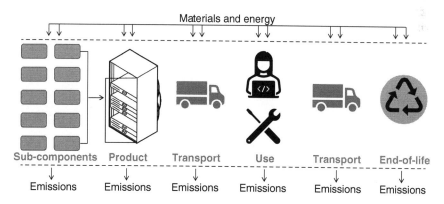

Figure 9.3 Product life cycle.

---

35 Flucker, S., Whitehead, B, Tozer, R. (2017). *Minimising data centre environmental impact – beyond energy efficiency*. CIBSE ASHRAE Technical Symposium 2017.

36 ISO 14040:2006 *Environmental management – Life cycle assessment – Principles and framework*.

37 ISO 14044:2006 *Environmental management – Life cycle assessment – Requirements and guidelines*.

results means that vendors are rarely able to guarantee that their products have been produced and disposed of in a way which does not include unethical business practices, breaches of human rights, damage to worker health, and environmental pollution.[38]

Worse, if only considering operational energy consumption, there is a risk of burden shift, e.g. by upgrading older, less efficient equipment to save energy. This saving will be at least partly offset by the increase in the embodied impact (the net impact could even be worse). A similar argument can be made for developing a new facility, which can employ new high-efficiency systems, versus upgrading a legacy facility, which may have limitations on the performance which can be achieved but will have a reduced embodied impact – it is not clear which is better. Without having the tools and data to measure the trade-offs, decisions are not made in an informed way.

There is a perception that refurbished hardware is less reliable than new and that newer hardware is more efficient. However, research indicates that employing professionally remanufactured servers can be more energy efficient than using the latest generation of servers if configured properly.[39] This can be attributed to the slowdown in Moore's law and the fact that newer servers are not maintaining the same efficiency improvements seen in the past. Reusing servers saves a significant amount of waste sent to landfill and reduced waste and toxic emissions produced during the manufacturing of servers.

Data centre embodied environmental impact can be of similar order or magnitude as the operational environmental impact.[40] Studies show that the IT equipment has a high embodied impact due to the materials and processes involved in producing electronics, its high refresh rate, and polluting disposal practices.[41] The embodied impact of the infrastructure (building services equipment and systems) is also relatively high compared to that of the building itself.

Part of the difficulty of addressing embodied impacts is quantifying them. There are thousands of parameters which can be measured, not just carbon emissions; these include water consumption, air and water pollution, global warming, damage to ecology. . . These impacts are grouped into damage categories, typically resource use, human health and ecosystem quality. In order to facilitate analysis of overall impact and trade-offs, a common unit is used, such

---

38 ICLEI Members (2020). *How to Procure Fair ICT Hardware. Criteria Set for Socially Responsible Public Procurement.* Freiburg, Germany: ICLEI – Local Governments for Sustainability, European Secretariat Leopoldring 3.

39 Bashroush, R., Rteil, N., Kenny, R. and Wynne, A. (2022). Optimizing server refresh cycles: The case for circular economy with an aging Moore's law. *IEEE Transactions on Sustainable Computing,* 7 (1), pp.189–200. doi:10.1109/tsusc.2020.3035234.

40 Whitehead, B., Andrews, D. & Shah, A. *The life cycle assessment of a UK data centre. Int J Life Cycle Assess* 20, 332–349 (2015). https://doi.org/10.1007/s11367-014-0838-7.

41 Shah, A., Bash, C., Sharma, R., Christian, T., Watson, B. and Patel, C. (2009). *The environmental footprint of data centers.* IPACK2009-89036.

as Eco-indicator points which equate to the impact of an average European during a year. In the case of a data centre, a functional unit of points/kW IT/year may be used. LCA is a resource-intensive process in part due to data availability. High variability of results can be achieved depending on assumptions, data used, chosen characterisation method (which turns an emission into an impact), and even the analysis software used. With the application of any metrics, there is a balance to strike between accuracy and ease of use.[42]

One way to reduce embodied impacts is dematerialisation, i.e. designing out unnecessary components. This has the added advantage that simpler designs tend to be more reliable, easier to operate, and have reduced capex and maintenance costs. Design for disassembly helps ensure that components of a product can be more easily removed and reused in refurbished/remanufactured products at the product end-of-life, thus enabling a more circular approach.[43] This is one of the objectives of the Open Compute Project.[44] The secondary market for materials is likely to grow due to pressures on the availability and price of raw materials.

There is increasing awareness that sustainability is not just energy efficiency. Although corporate social responsibility and brand value are driving change, the business case for addressing embodied impacts is less tangible.

Green Building Certifications

Some data centres may be independently accredited in order to demonstrate their green credentials. Schemes include BREEAM,[45] LEED,[46] and NABERS.[47] These originate from green building certifications from other building types such as offices and are based on achieving specific credits in order to reach a scoring level. Although they use established and auditable methodologies, in most cases, the scoring does not reflect the holistic data centre environmental impact and focuses on their energy efficiency, rather than the significant embodied impacts.

42 Tozer, R., Flucker, S., Whitehead, B., Andrews, D, Summers, J. (2018). *Data Center Sustainability Index.* ASHRAE.

43 Kerwin, K., Andrews, D., Whitehead, B., Adibi, N., Lavandeira, S., *The significance of product design in the circular economy: A sustainable approach to the design of data centre equipment as demonstrated via the CEDaCI design case study.* (2022) Materials Today: Proceedings, Volume 64, Part 3, Pages 1283-1289, ISSN 2214-7853, https://doi.org/10.1016/j.matpr.2022.04.105

44 Open Compute Project. *Sustainability.* [online] Available at: https://www.opencompute.org/projects/sustainability [Accessed 6 Nov. 2022].

45 BRE Group (2022). *BREEAM – BRE group.* Available at: https://bregroup.com/products/breeam/.

46 USGBC (2020). *LEED rating system.* Available at: https://www.usgbc.org/leed.

47 tessabreakspear (2018). *Data Centres | NABERS.* [online] Nabers.gov.au. Available at: https://www.nabers.gov.au/ratings/spaces-we-rate/data-centres.

Policy and Regulation

The high environmental impact of data centres has not escaped the notice of policymakers. The EBC's *International review of energy efficiency in data centres for IEA EBC Building Energy Codes Working Group*[48] presents a review of international policies and standards relating to data centre energy efficiency including voluntary schemes and suggests possible future policies.

One way of influencing the market is through green procurement practices, i.e. actively seeking or requiring more sustainable solutions. The European Commission has developed green public procurement (GPP) criteria for different areas including data centres, in recognition of the fact that Europe's public authorities are major consumers and so can influence the market for goods and services.[49]

The World Green Building Council's *Net Zero Carbon Buildings Commitment*[50] calls on businesses, organisations, cities, and subnational governments to reduce (and compensate where necessary) all operational and embodied carbon emissions within their portfolios by 2030, and to advocate for all buildings to be net zero whole life carbon by 2050. Currently commitments made by governments to date fall far short of what is required, with national climate plans projected to increase global greenhouse gas emissions by almost 11% by 2030, compared to 2010 levels.[51]

It is important that policy also takes a holistic view of environmental impact and does not promote perverse incentives, e.g. support for waste heat recovery that does not incentivise reducing the amount of heat produced through energy efficiency.

Much of the work to improve the sustainability of data centres is driven by economics (saving energy saves operating costs and increases profitability and competitiveness) as well as corporate social responsibility pressures. However, there is also a concern that unless the industry acts voluntarily, it will face increasing legislative restrictions.

Conclusion

Sustainability needs to be embedded into all aspects of data centre design, build, and operation. It is not enough just to buy renewable energy or design with a low PUE – action is required by all stakeholders throughout the value chain. In order to make a real impact, an understanding of key areas to prioritise is important – not just making easy token gestures.

---

48 (2022). *International Review of Energy Efficiency in Data Centres for IEA EBC Building Energy Codes Working Group*. Brocklehurst: Pacific Northwest National Laboratory.

49 European Commission, Joint Research Centre, Dodd, N., Alfieri, F., Gama Caldas, M., et al., *Development of the EU Green Public Procurement (GPP) criteria for data centres, server rooms and cloud services*: final technical report, Publications Office, (2020), https://data.europa.eu/doi/10.2760/964841

50 World Green Building Council. (n.d.). The Commitment. [online] Available at: https://worldgbc.org/thecommitment/.

51 United Nations (2022). *Net Zero Coalition*. [online] United Nations. Available at: https://www.un.org/en/climatechange/net-zero-coalition.

10

The Importance of Planning to Avoid Things Going Wrong

Introduction

Data centres are critical infrastructure helping to enable people and communities to function and thrive. The exponential uptake during the lockdown in subscription streaming services such as Netflix or Spotify and acceleration in the use of communication platforms such as Zoom and Microsoft Teams has increased average bandwidth usage per household and, in business, the increased monetisation of data and growth in cloud services are driving ever-growing demand for data centre capacity.

This demand has fuelled industry restructuring with increased private equity and infrastructure fund investment and accelerating M&A activity. The world's increasing, insatiable demand for data, online content, digital communication, and entertainment will be dependent on the ability to provide further data centre capacity and is resulting in:

- disruption in the capacity market for hyper-scale data centres and an increased trend for users to bring critical data services 'local' (i.e. ever closer to data end users) to build resilience, reduce reliance on global network bottlenecks and, ultimately, increase the speed of access for data users including 'edge' data centres.
- requirement for reliable power sources driving an increased need for electricity supply for robust energy storage solutions.
- increased political, regulatory, and reputational pressure to decarbonise, affecting all elements of both new developments and existing operations, including:
  - renewable energy supply and heat offtake
  - reporting on carbon usage throughout global supply chains.

Data Centre Essentials: Design, Construction, and Operation of Data Centres for the Non-expert, First Edition. Vincent Fogarty and Sophia Flucker.
© 2023 John Wiley & Sons Ltd. Published 2023 by John Wiley & Sons Ltd.

- an ever-increasing focus on the value of data as an economic asset requiring all data centre businesses to address with care data-related issues including cyber security, liability risks, and regulatory compliance.
- a need for increased collaboration and the use of a variety of traditional and more innovative business models to support the development of new capacity, internationally and locally, supported by policy.
- increased competition for assets as the industry matures and the fight for scale accelerates. Whilst hyper-scalers have expanded their market share and continue to invest and build across the world, the co-location market (led by larger players such as Digital Realty and Equinix) has seen a lot of M&A activity. This trend extends to the mid-market with a number of private equity investors and infrastructure funds investing in a range of co-location businesses and embarking on 'buy and build' programmes both in the United Kingdom and across Europe.

As this brief overview indicates, when we talk about legal issues affecting data centres, we are talking about many different topics, covering financing, joint-venturing, planning and consenting, construction procurement, supply chain management, value-adding services, data management, health and safety, employee management, defects and commercial claims, and everything in between.

The common thread that runs through these topics is the industry's need to bring new capacity to the market, quickly and effectively, in compliance with local law and regulatory requirements and to deliver strong business models to ensure the data economy can thrive for the benefit of our communities.

In legal terms, this may require knowledge of areas as wide-ranging as network and energy strategy, planning and consenting issues, the development of district heating solutions, all aspects (regulation, funding, contracting, and servicing) of the implementation of greener energy solutions (including renewables, heat, storage, and electricity), the navigation of energy regulation, and the negotiation of favourable power purchase agreements.

This chapter serves as an introduction to various legal issues that may arise in connection with data centres. With this in mind, a useful place to start is to look at the legal issues that any potential acquirer or investor in a data centre provider will want to consider in the context of its legal due diligence. Following that, we look at some of the major themes in a little more detail.

Acquisitions and Investments

The market for data centre transactions continues to grow apace, with increasing deal values and volumes.

In the last couple of years alone, we have seen:

- the acquisition by investment firm KKR and infrastructure investor Global Investment Partners of CyrusOne Inc., a global REIT specialising in the design and construction of more than 50 high-performance data centres worldwide. This all-cash transaction was valued at approximately US$15 billion[1];
- the announcement of an agreement to acquire Switch, Inc.,[2] a leader in exascale data centre ecosystems, edge data centre designs, industry-leading telecommunications solutions and next-generation technology innovation, by Digital Bridge Group, Inc., and IFM Investors for approximately US$11 billion;
- American Tower Corporation's acquisition of data centre developer, CoreSite Realty Corporation[3] for US$10.1 billion; and
- the acquisition of QTS Realty Trust by various Blackstone funds[4] for approximately US$10 billion.

These transactions display an interesting mix between financial investors and trade buyers and – within the financial community – a mix of investors with a track record in data and centres and those who are new to the market.

Additionally, we have seen new structures emerging, at least in part because of the intensive demands for capital expenditure that data centres demand. Examples of this include Equinix's $1bn joint venture with GIC, Singapore's sovereign wealth fund, and KKR committing $1bn to a new European platform, Global Technical Realty, to build new capacity alongside an ambitious 'roll-up' acquisition strategy.

Any investor or acquirer will want to see certain key legal issues addressed in its legal due diligence. Before turning to those, it is worth noting that the legal due diligence is only one part of the due diligence that an acquirer will likely carry out. Other areas on which they will focus include:

- physical infrastructure: the topology of power and cooling systems (which feed into the 'Tier' ratings – see further 'Ensuring Resilience' later) and standards to which the facility has been maintained.
- commercial: due diligence would cover the data centre market and the operator's competitors, along with the data centre's siting, ongoing operations, and any issues impacting its operation.

---

1 KKR and GIP Complete Acquisition of CyrusOne – CyrusOne.
2 Switch to be Taken Private by DigitalBridge Investment Management and IFM in $11 Billion Transaction – Switch, Inc.
3 267527_002_BMK_WEB.PDF (gcs-web.com).
4 QTS Realty Trust to Be Acquired by Blackstone Funds in $10 Billion Transaction – Blackstone.

- financial and tax: the acquirer's accountants will want to look at the historical accounts of the business as well as potential synergies going forward.

Additional due diligence may focus on IT, environmental aspects, and insurance.

Legal due diligence essentially serves two purposes: first, the gathering of information relating to the business, and second, the identification of specific issues that may impact upon the price offered or that may require attention in the sale documentation to be negotiated between the acquirer and the seller or sellers. In particular, where a risk is identified, the sale documentation can allocate risk between the acquirer and seller. An investor or acquirer will want to look at the following areas:

- **Corporate structure and finance:** first, an acquirer or investor will want to understand the structure of what it is acquiring or investing into. As outlined earlier, we have recently seen a greater degree of variation in this respect than has historically been the case. The nature and structure of any existing financing will also be of key importance. An acquirer will likely be looking to replace that financing and any existing security over the target's business and assets will need to be released prior to acquisition.
- **Assets:** the extent of the assets owned by the data centre operator will depend upon the relevant operating model which it has chosen to adopt. An acquirer or investor will want to know, however, that the data centre operator owns or has a valid right to use, each of the assets that it uses to run its business.
- **Insurance:** an acquirer or investor will want comfort as to the extent of insurance coverage that the operator has in place and that the scope and level of insurance are in line with market norms. Also of interest will be the operator's loss history and details of any ongoing insurance claims and any consequent impact on future insurance premia or the ability to obtain insurance on similar terms.
- **Material contracts:** one of the key concerns of a financial investor will be the robustness of the data centre's income streams. The key customer contracts will need to be reviewed – looking at term and early termination rights, in particular, to determine how robust the associated revenue is. The potential liabilities under those contracts will also be of concern, along with the supply contracts that will need to be in place for the data centre operator to be able to operate the data centre and meet its obligations towards its customers.
- **Litigation:** any ongoing or threatened litigation may have significant reputational and/or financial implications for an investor. Litigation can take many forms and may involve a supplier (e.g. a failure to provide power or construction delays (see real estate later)), a customer if KPIs consistently fail to be met or third parties if, for instance, there is an allegation that third party intellectual property is being infringed in some way by the data centre operator. The risks

attached to any such litigation will likely be apportioned in the acquisition or investment documentation.

- **Compliance with regulations:** an acquirer will want details of all licences, consents, and registrations required or obtained by the data centre operator in connection with the operation of its business – and potential breaches of any such licences, etc. An acquirer will also want to know about any investigation, enquiry, or prosecution by any governmental or administrative body into the affairs or operations of the operator. Details of anti-corruption policies and procedures implemented by the operator to ensure compliance with bribery legislation will also be requested. For more detail on regulatory compliance – particularly in respect of data and cyber – please see the later section.

- **Intellectual property rights:** intellectual property rights (IPRs) are an important consideration for any data centre acquisition. A financial investor should consider who owns the data centre's resources and how others may use them. For more detail on this, see the later section.

- **Information technology:** an investor will want to know that the operator's IT assets are fit for purpose and regularly and appropriately maintained, and that there have been no major system outages in the recent past. The continuation of any licences for the use of systems beyond the proposed transaction will also be checked.

- **Competition:** an acquirer or investor will want to know if a target business or joint venture party is or has been involved in any competition law infringement, or whether its business practices could give rise to future competition law investigations or claims. Appropriate warranties and/or indemnities in the sale documentation should be negotiated to address and allocate such risk. Additionally, appropriate competition law compliance measures (e.g. use of 'clean teams') should be implemented during the due diligence and transaction process to minimise the risk of sharing competitively sensitive information between transaction parties, particularly where the parties are actual or potential competitors.

- **Employees:** depending on the structure of the operator, contact with management may be a key element of the transaction. An acquirer will want to review the employment contracts of key personnel to check notice periods and – if considered sufficiently important – may wish to put new contracts in place with key personnel to ensure employee retention. Data protection rules will determine how much data (and in what form that data) can be disclosed to investors.

- **Real estate:** an acquirer will want details of the basis on which the land on which the data centre is sited is held by the operator, whether freehold or leasehold. Rights of access will be of key importance, and an acquirer will also want details of any other rights benefiting the property along with any third-party rights to which the property is subject (including restrictions and covenants which may be onerous or unusual, or which conflict with current use).

- **Environmental:** an acquirer will want access to any environmental reports, audits, or other assessments carried out in the past few years. It will also want to know if the operator is or may be in breach of any environmental law.
- **Health and safety:** documents underpinning the operator's health and safety system should be made available. An investor will want to be comfortable with the robustness of such a system and also to know of any notifications or complaints, and any current or threatened legal proceedings.

An investor or acquirer will also want to investigate to what extent a potential transaction may give rise to merger control and/or national security issues.

- **Merger control:** merger control involves a forward-looking assessment, by a competition authority, of whether a transaction is likely to result in a substantial lessening of competition in any relevant market and, if so, it could be approved subject to certain legally binding conditions, or prohibited altogether. An acquirer or investor will need to assess at an early stage before the drafting of the sale documentation is finalised, whether the proposed transaction may be subject to merger control review and approval in any geographic jurisdiction. This will typically require gathering and analysing information about the local and international turnover, market shares, business activity, assets, and corporate structure, of each party involved to determine if relevant merger review thresholds are triggered. Depending on the applicable regime, merger notification, and review may be mandatory and suspensory (the transaction must be cleared by a competition authority before completion, with no business integration permitted before the merger is cleared on pain of substantial fines and potential divestment orders) or voluntary (completion may legally occur before approval is granted by a competition authority, but with the risk of the transaction potentially having to be modified or unwound if a subsequent merger review identifies serious competition concerns). Appropriate conditions precedent should therefore be negotiated in the sale documentation that address merger control.
- **National security and foreign investment control:** an acquirer or investor will need to assess at an early stage, again, before drafting of the sale documentation is finalised, whether the proposed transaction may be subject to review and approval by a government body on the basis of national security or public interest considerations. This will typically require gathering and analysing information about the industry sectors, business activity, assets, corporate structure, ownership, and any relationships with government bodies, of each party involved in the transaction, but specific requirements vary depending on the particular regime (including whether notification and approval requirements are mandatory or voluntary; and whether approval must be obtained before completion or retrospectively). Depending on the applicable regime,

completing a transaction without prior approval can result in the transaction being deemed void, and expose the acquirer (and potentially its senior management) to substantial civil and/or criminal sanctions. The regime may apply to national (not just foreign) investors, as well as foreign-to-foreign transactions that have a 'nexus' to a particular geographic jurisdiction. For example, in the UK acquisitions or investments involving 'data infrastructure,' including data centres which carry out certain specific activities, are subject to prior mandatory notification requirements if the acquisition of shares or voting rights meets prescribed thresholds. Transactions involving data centres that do not trigger the mandatory thresholds, or which only involve assets, may nonetheless be subject to review by the UK government under its 'call-in' power. Appropriate conditions precedent and warranties should therefore be negotiated in the sale documentation that address national security and foreign investment control.

Operating Models and Commercial Contracts

Expanding businesses must choose whether to build or lease data centre space (either directly, through a co-location or managed services model, or indirectly through cloud-based services). A balance must be struck between the desire for control, on the one hand, and the appetite for taking on associated operating and capital expenditure, on the other – but always bearing in mind the future plans and requirements of the business.

The business models supporting data centres range from owner-operator or single-tenant models through to colocation providers, fully managed hosting, and cloud-based services. Most of these models require a different approach to commercial contracting.

Contracting for data centre services should always align with the specific service requirements, with relevant services clearly documented and a pricing schedule calibrated to the scope and level of services:

- Customers need to consider their requirements at the outset, including whether service levels need to guarantee a 'gold-plated' service (e.g. regarding availability and downtime) or whether a lesser standard is acceptable for a lower price.
- Each operator needs to consider what service levels it can guarantee, the boundaries of its control and responsibility, and appropriate exceptions/relief to performance regimes (e.g. planned downtime and customer responsibilities).

The contract will need to apportion rights, risks, and obligations in relation to all aspects of the service, including data migration and testing, security (both physical and cyber) and access rights, and build in sufficient flexibility to accommodate changes. The performance regime and associated liability will need special care and may incorporate escalating remedies such as enhanced monitoring, step-in rights and, ultimately, termination and exit.

Funding Investment

New funding sources are being pursued by developers – private equity, strategic investors, and banks. Limited recourse finance structures are attracting specialised infrastructure investors who see opportunities for sustainable returns from the data centre market. Where the owner of a data centre is also the owner of the land on which the data centre sits, some form of charge over that land may form part of the security package that lenders seek. Other operators have formed joint ventures and leveraged their capital via off-balance-sheet financing or adopted an OpCo/PropCo structure to reduce tax exposure. All investors will focus on the offtake contracts (term/quality of offtaker and performance security) and the ability to support sustainable returns.

Construction

Not unexpectedly, data centre procurement and construction considerations vary depending on location, scale, complexity, and market demand.

Choice of Contract

Data centre construction combines relatively simple civil/structural elements on the one hand with complex integrated supply and instals requirements on the other, with hyper-scale data centres leaning more towards developer-driven contracting strategies (as would be usual for large-scale greenfield projects in any industry) and smaller containerised or modular solutions being supply-side driven (with bespoke procurement terms more typical of off-the-shelf OEM offerings). It is also not unusual to see a mix of these approaches on projects where modular solutions are used to develop large-scale offerings with associated building, energy, and connectivity infrastructure provided by other contractors/suppliers.

There is no industry-targeted standard form or tailor-made contracting solution available and data centre construction often requires a combination of approaches (ranging from construction only to design and supply arrangements) on the same project.

FIDIC is a popular global form which is flexible enough to allow for a variety of procurement arrangements from the same international standard form contracting base, but local forms (such as JCT in the United Kingdom) also find favour.

The NEC form is popular in geographies where the supply chain is local, and the NEC form is well-known and widely used. Globally, however, that tends to be the exception rather than the rule, with construction requiring significant imported content from a supply chain which is geographically widespread. Whatever the standard form used as the base document, there is a need to include suitable modifications to deal with industry-specific risks.

International or widely used standard forms such as FIDIC and the NEC are often perceived to be too contract management-intensive for smaller projects. However, care should be taken not to take a blanket approach to this as FIDIC, for example, has various alternative contracts ranging from the 1999 to the 2017 Red (suitable for Construction to Employer's design), Yellow (design build) and Silver (full turnkey/EPC) Book editions, to the more simple Green Book (tailored to less complex, shorter duration and/or lower value projects).

The benefit of using standard forms, even though they require modification, is that they are well known across a broad geographical footprint and widely accepted (sometimes translated into languages the supply chain is familiar with) and suitable for varying legal frameworks (Common Law, Civil Code, etc.), thereby reducing unnecessary negotiation time and effort.

Technical and Pricing Schedules

Whatever the form of contract, the technical and pricing/payment profile documents need to be appropriately drafted to suit. Not only do these documents need to deal comprehensively with the works by the contractor/supplier but also the design, off-site and on-site requirements, and assumptions for integration with works provided by the developer or others. One should never rely on the contract terms to fill technical gaps.

Risk Allocation

Data centre construction is not immune to the usual tension between developers driving, and suppliers seeking to avoid, fitness-for-purpose obligations and single-point responsibility. That tension plays out in many industries and the data centre market is not unique in that regard.

However, it is worth noting that in the construction of data centres, the integration risk (especially around connectivity, energy, and climatic conditions) and the responsibility for a fit-for-purpose solution which is certification ready (as opposed to certification being a prerequisite for takeover) together with the significant risk of damages, including indirect or consequential loss (see further 'Disputes' in the later section) presents a unique convergence of risk which contractors and suppliers are not prepared to accept or cannot accept by virtue of their inability to flow down that risk to crucial elements of their supply chains.

Data centre procurement and construction contracting require risk allocation to be proactively considered at the procurement stage with collaborative engagement and transparent discussions with the supply chain, so that risk and responsibility are understood, transparently allocated, and proactively managed. This is the key driver to avoiding unnecessary issues and ultimately disputes.

Usual Suspects

Typical matters which are heavily negotiated include:

- rise and fall in input costs (inflation) across a broad range of commodities;
- shipping and transport availability, as well as the availability of docking and transport routes in more remote locations;
- local content obligations and responsibility for local contractors/suppliers;
- local and foreign taxes;
- imported content and custom duty liability;
- performance and warranty bonds;
- IPRs and resultant liability for infringement;
- site delays, including delays to services and infrastructure provided by others;
- the definition of appropriate access;
- fitness for purpose;
- takeover/performance testing requirements and liability for tier ranking or other certification, as a prerequisite for takeover;
- liability for defects (extent and duration, including liability for latent defects);
- liability for supply chain insolvency;
- liability for performance failures;
- liability for performance specifications and verification of the developer's data/requirements; and
- quantum and extent of liquidated damages for delay and performance, linked to overall limitation of liability and exclusion/liability for indirect and consequential losses.

If these matters are not appropriately addressed, they will inevitably lead to disputes together with the usual construction family favourites of delay and disruption and variations.

Infrastructure Provision and Project Rights

It is vital to consider the essential infrastructure that will support the operation of the data centre, including energy, cooling systems, and data connectivity (see further Chapter 5). Whether from a supplier perspective at the early stages of development, or for a customer looking for cost efficiencies, the enabling infrastructure can make a huge difference to the economics of the deal. At one end of the scale, it may be preferable to host or procure data centre services closer to large population centres with robust pre-existing telecom infrastructure. At the other extreme, there may be energy efficiencies to be capitalised upon by locating data centres close to sources of geothermal energy, stable renewable energy sources, or naturally cooling climatic conditions. The type of services to be provided or procured (e.g. short-/long-term storage, critical/business-as-usual functions) will be material in driving decision-making on these issues.

Understanding the consenting structure and timetable is of key importance.

A key element of any data centre development will be to obtain the relevant project rights, which are essential to obtaining debt funding:

- **Energy supply:** with the clock ticking on net zero commitments, organisations are committing to de-carbonising their energy supply chains, using Green PPAs and/or on-site generation. This is particularly important for data centres given their high energy requirement. There is also scope to sell exhaust heat to district heat providers for auxiliary heat loads. This is often overlooked but should feed into the site selection criteria, particularly where sites are close to the built environment (see further Chapter 3).
- **Grid connection:** data centres require significant energy-loading capacity. The timing, negotiation, and development of the solution require significant discussion early on with transmission or distribution network operators and/or independent connection providers to ensure timely delivery and may also impact the siting of the data centre due to locational grid constraints.
- **Property rights:** the property strategy needs to feed into the site-selection criteria, alongside the energy solution. Early advice should be sought to secure exclusivity to the land in order to progress to a long-term lease or acquisition upon achievement of planning consent and grid rights. Achieving certainty in these terms early on is key to the project rights strategy.
- **Planning and consenting:** the key issues to understand when considering planning and consenting for data centres, particularly in Europe and the Middle East, are these:
  - Some countries and cities have zonal regimes which permit data centres as part of business uses, whereas others have permit-based systems.

- Some countries treat data centres as industrial land uses, while others treat them as business or storage uses. This makes a difference in zoning site suitability and availability and permit applications.
- Federal vs. state planning requirements differ in a number of countries.
- Energy security of supply (including backup generators – sometimes gas or diesel), energy regulation, energy efficiency, and carbon footprint issues are increasingly an issue in applications for planning permits, along with footprint land take, security of energy supply, the need for high security, and the fact that there are relatively few on-site jobs. This often necessitates economic growth and supply chain 'need cases' to be produced as well as carbon reduction and minimisation cases.
- Noise can be a material planning consent issue and appropriate mitigation approaches are key.

Construction

Construction risks include defects, supply chain insolvency, performance specification failures, and local market volatility. Delays mean lost revenue and performance failures during operation may give rise to contractual penalties as well as lost revenue. Mitigation requires upfront planning and a careful assessment of the best contracting model, to achieve a balance between speed of delivery, price, and quality – from separate work 'packages' through to combined EPC 'turnkey' models. Developers must be realistic about the resource available to manage the works and understand applicable regulatory standards and market forces. Efficiency-increasing techniques such as off-site fabrication and modular construction are coming increasingly into play. Construction risk should be identified and proactively managed from the outset. See further Chapter 6.

Moving to Low Carbon Solutions

In the United Kingdom, National Grid ESO (the operator of the transmission network) expects data centres to consume just under 6% of the United Kingdom's total energy consumption by 2030.[5] In Denmark, data centre energy use is expected to triple by 2025 and account for 7% of electricity demand.[6] And

---

5 download (nationalgrideso.com).
6 Data Centres and Data Transmission Networks – Analysis – IEA.

in Ireland, data centre electricity use was already at 14% of total energy consumption in 2021.[7]

With our increasing desire for instant data, growth in 5G, the Internet of Things, the metaverse, machine learning, bitcoin, and blockchain, demand for data processing and the need for data centres are only going to increase. Such huge power demand from one sector of the market will not only have a significant impact on grid capacity, but also sustainability of the sector (unless low-carbon and energy-efficient solutions are implemented across the board).

What Are Data Centre Operators Doing to Transition to Low Carbon?

In Europe, cloud and data centre operators have signed up to the Climate Neutral Data Centre Pact,[8] a movement to eliminate the environmental footprint of data centres by 2030. Most of the major players, including AWS, Google, and Microsoft, are signatories. The self-regulatory initiative sets goals in energy efficiency, clean energy, water usage, circular economy, and circular energy systems. From January 2023, signatories will be expected to certify adherence every four years.

To give a few examples of the goals that have been set: (i) data centre electricity demand will be matched by 75% renewable energy or hourly carbon-free energy by the end of 2025 and 100% by the end of 2030; (ii) from 2025, new data centres operating in cool climates will meet an annual power use effectiveness (PUE) target of 1.3 and 1.4 in warm climates; and (iii) data centre operators will explore possibilities to interconnect with district heating systems and other users of heat to determine if opportunities to feed captured heat from new data centres into nearby systems are practical, environmentally sound, and cost-effective.

As the Pact is voluntary at the moment, there are no penalties for failing to achieve the published goals. The European Commission will monitor performance alongside data centre trade associations, so it is possible that the goals will become enshrined in regulation over the next decade. The key drivers at the moment are likely to be ICT Companies' own CSR and ESG policies and commitments, customer and funder requirements and public expectations. In a recent survey, 30% of customers wanted demanding and contractually binding efficiency and sustainability commitments, with a further 44% expecting at least some efficiency and sustainability commitments in their contracts.[9] See further Chapter 9.

A lot of the tech giants are also signatories to RE100, a global initiative bringing together the world's most influential businesses committed to 100% renewable

---

7 Data Centres and Data Transmission Networks – Analysis – IEA.
8 Climate Neutral Data Centre Pact – The Green Deal need Green Infrastructure.
9 Data Center Frontier, 2021.

power. ICT companies currently account for the majority of all corporate power purchase agreements, a mechanism for buying renewable energy directly from a renewable generator. By 2022, Amazon, Google, Meta, and Microsoft had contracted for over 38 GW, from a combination of solar and wind generators.[10]

There are also various European and international voluntary standards. CENELEC, the European Committee for Electrotechnical Standardization, the European Union's principal electrical standards authority, developed the EN 50600[11] series of standards for data centre facilities and infrastructure. ISO (the International Organization for Standardization) has published the ISO/IEC 22237 series[12] which specifies requirements and recommendations for the parties involved in the design, planning, installation, operation, etc., of data centres and the ISO/ICE 30134 series[13] which specifies KPIs for data centre operation. These standards include flexible requirements in relation to power distribution, renewable energy usage, environmental control, energy, heat and water effectiveness, and reuse. Although currently voluntary, we would expect these standards to be drawn on to form the basis of any European or International regulation going forward.

Corporate Power Purchase Agreements (CPPAs)

Data centres, especially those operated by global ICT companies, are attractive to renewable generators because they tend to have good covenant strength and require substantial amounts of power. A generator would much rather contract all of its output to one or two off-takers, rather than having to negotiate five or six CPPAs with different counterparties. CPPAs are complex documents, and the larger ones can take from six to nine months to negotiate. Running numerous negotiations would be extremely costly.

The issue for data centre operators is that data centres require 24/7 baseload power. Renewables by their nature are intermittent and generation output is variable, depending on how much the sun is shining, how strong the wind is blowing and from what direction, and whether all equipment is functioning correctly. So even if a data centre operator contracts for 100 MW of renewables for a 100 MW data centre, he will still need to purchase and sell additional electricity to smooth

---

10 Data Centres and Data Transmission Networks – Analysis – IEA.

11 Explaining the new family of ISO Data Centre Standards – Techerati.

12 ISO/IEC 22237-3:2021(en), Information technology – Data centre facilities and infrastructures – Part 3: Power distribution.

13 ISO/IEC 30134-9:2022(en), Information technology – Data centres key performance indicators – Part 9: Water usage effectiveness (WUE).

out the peaks and troughs of renewable generation. Currently, this is likely to mean buying non-renewable power in the wholesale market.

On-site Generation and Electricity Supply

Data centres tend to be built in urban areas with limited space and grid capacity for large-scale on-site renewables. It may be possible to put 1 MW plus of solar power on the roof of a data centre (depending on the roof area and direction it faces), but this would only be a tiny proportion of the data centre's needs. There is also the option of purchasing power from a neighbouring solar farm or onshore wind farm through a private wire, but the opportunities are limited by location. For these reasons, CPPAs are a more attractive and obtainable solution for data centres, although that is not to say that a mixture of on-site and off-site renewables is not possible.

Local regulatory requirements for generating and supplying electricity will need to be given careful consideration, especially for co-located data centres (with multiple occupants), when deciding on power procurement for a site. In the United Kingdom, the Electricity Act, 1989 makes it an offence to generate, distribute, or supply electricity without a licence or an exemption.[14] The Electricity (Class Exemptions from the Requirement for a Licence) Order, 2001 (the 'Exemptions Regulations') provides various exemptions for the generation, distribution, and supply of electricity.

The small generator exemption would typically apply to most on-site generation, as it allows generation of up to 10 MW (or 50 MW if the generating station has a declared net capacity below 100 MW) without a licence.

There is a blanket distribution exemption for operators distributing electricity to non-domestic premises, which would usually apply to a data centre operator. A data centre operator would be considered to be distributing electricity to the extent it operated electrical wires on its site to convey electricity, such as a private wire from on-site generation or wiring from the grid supply point to co-location tenants.

The supply exemptions are more complex and scenario dependent and legal advice should be sought before relying on these. If a data centre operator were to purchase renewable electricity on behalf of multiple co-location tenants within a data centre via a CPPA, it should be able to re-sell that power to those tenants without a licence,[15] as long as that power had been supplied directly to the data centre operator by a licensed electricity supplier. This would typically be the case

---

14 ss4(1)(a), (bb), (c) and 5(1) of the Electricity Act 1989.
15 Schedule 4 of the Exemptions Regulations, Class B (Resale).

with a CPPA as the data centre operator would also need to contract with a licensed supplier for the CPPA electricity to be supplied to the premises.

More tricky is the scenario where the data centre operator wants to supply both renewable power purchased via a CPPA and power generated on-site or near-site (whether renewable or backup). The on-site supply exemption[16] restricts the supply of electricity to multiple unconnected consumers (as would be the case for a co-location data centre) to 100 MW. There are also tests to be met for the generating station to be considered to be on the same site as the data centre (immediately adjacent, or separated only by a road, railway, or watercourse) and restrictions as to who can own the private wires over which the electricity is conveyed to the consumers. It is also worth noting that if the data centre operator were to be supplied with on-site power generated by a third party, the data centre operator would not then be able to resell that electricity to its tenants.

Backup Power

With a requirement to run 24/7, 365 days a year, data centre operators need to be certain that they will always have sufficient power. This means that they will all have backup generation on-site in case the grid fails for any reason. As mentioned earlier, on-site renewable generation is unlikely to have sufficient capacity to be used as backup generation (and also does not provide sufficient baseload power). Backup generation is typically diesel generation, which is certainly not low in carbon. Arguably, backup generation is never or rarely used, so it does not contribute to carbon emissions in the same way as baseload power. However, the very fact that diesel is purchased and stored by data centre operators contributes to the continuing of the diesel market.

What are the alternatives? On-site battery storage? As per on-site renewables, an on-site battery would also require sufficient land to match the data centre's power demand. This could be tricky in urban areas. 100 MW batteries are big, expensive batteries and would potentially need to be operated by a specialist third party to achieve the revenue stack required to repay the capex. A battery of this size could be used to help alleviate constraints on the grid when it was not being used for backup. One issue is keeping sufficient charge in the battery so that it can respond immediately to provide power to the data centre when the grid goes down, versus the flexibility required to buy and sell power in the market to alleviate grid constraints and achieve the desired revenue stack. The other issue is the one mentioned earlier in terms of licence exemptions. Batteries are treated as generation assets. A large battery is likely to need a generation licence, and the

---

16 Schedule 4 of the Exemptions Regulations, Class C (On-site supply).

complexities of the supply exemptions arrangements may not allow sufficient power to be supplied to the data centre through a private wire.

Some data centre operators have been testing hydrogen-powered fuel cells for backup power. Hydrogen fuel cells combine hydrogen and oxygen to produce electricity, heat, and water and as such are considered more sustainable than diesel generators which emit air pollutants and greenhouse gas emissions. In the summer of 2022, Microsoft tested a 3 MW hydrogen fuel system, powerful enough to power around 10,000 computer servers.[17] However, the hydrogen that was used to power the fuel cell was not green – in order to be considered green, hydrogen needs to be produced from renewable energy. This causes logistical and financial challenges as a data centre operator would either need to have access to sufficient local or on-site renewable generation and sufficient land for an electrolyser or would need to purchase and transport green hydrogen produced off-site, which in itself has carbon implications. The market is not yet ready to commercialise large-scale green hydrogen fuel cells, but hopefully, we will see the market develop in the not-too-distant future.

Where Could Other Efficiencies Be Made?

As National Grid ESO points out, data centres operate 24/7 to allow for instant data processing, but this can lead to wasted energy consumption whilst systems are idling, waiting for the next surge of traffic, with only 10% of energy consumption being used for heavy computation work. If data centres were able to shift power demand to align more neatly with the generation profiles of renewable and low carbon solutions, this would go a long way to help improve their sustainability. However, this seems unlikely at present as a change in data consumption would need to come first. National Grid ESO is hoping to incentivise the development of baseload low carbon solutions (low carbon solutions that have a more stable generation profile) to assist with the transition to a net zero power grid. Such technologies would be hugely beneficial for decarbonising large baseload consumers like data centre operators.

Data centre operators have been making energy efficiency gains through improved efficiency of IT equipment and data centre facilities (e.g. chip design, power infrastructure, and cooling systems), largely led by the hyper-scale cloud data centre operators.[18] With the push towards decarbonisation of heat, there is a growing opportunity for waste heat producers to sell waste heat to heating

---

17 Can hydrogen fuel cells power Microsoft data centers? – The Verge.

18 2021.11.17-Green-Giants-White-Paper-Final.pdf (pioneerpoint.com).

network operators, increasing the sustainability credentials of both the data centre and the heat network.

As we approach net zero commitment dates, embodied carbon and Scope 3 emissions are getting more attention – these are the carbon and other greenhouse gas emissions that a company is indirectly responsible for via its supply chain (including in materials, manufacturing processing, construction, transport, etc). In the past, businesses (and consumers) considered it sufficient for a business to look at their operational carbon footprint only, but now the focus is shifting to the supply chain and consumers' use of products. For many businesses, Scope 3 emissions account for more than 70% of their carbon footprint.[19] When considering a data centre, there will have been significant carbon emissions in the manufacturing of the IT equipment (including the extraction and processing of raw materials), construction of the data centre, transport of equipment to the site, etc. Net zero cannot be achieved without addressing embodied carbon and Scope 3 emissions. To do this, businesses (including data centres) will need to re-evaluate their relationships with their supply chain and ensure that appropriate processes are put in place to incentivise the reduction of embodied carbon and Scope 3 emissions.

Ensuring Resilience

Ensuring the absolute minimum of downtime and consequential revenue losses is a key consideration in the design and operation of data centres, driving not only the inclusion of low latency backup systems for power, water, communications, and security but also physical location. A further measure of resilience is a reliable, highly trained, and fast-reacting operation and maintenance capability, able to provide a 24/7/365 service. All of these are factored into evaluation against the standards of a 'tier' ranking system for data centres, with high rankings becoming more and more popular amongst end users.

Intellectual Property Rights

As mentioned earlier, IPRs are an important consideration for the owning and operating of any data centre. Owners, operators, and investors should consider who owns the data centre's resources and how others may use them.

---

19 https://www2.deloitte.com/uk/en/focus/climate-change/zero-in-on-scope-1-2-and-3-emissions.html.

There are two components to this. First, the data that is collected, assembled, or generated. Second is the data system in which the data is stored and managed. The financial investor must establish the ownership and usage rights needed across the asset lifecycle are in place which includes reviewing licence terms for sustainability and transferability of the IPR in both components. Establishing reliable access to the data post-acquisition is key.

The data system component will consist of hardware and software elements. Due diligence will need to be undertaken that the necessary IPRs have been granted for the data system to run as envisaged after acquisition.

There is no overarching legislative framework that governs the ownership of data. However, there are certain overlapping legal rights that may impact the use of non-personal data. In addition to contractual rights, the financial investor will need to consider confidentiality and trade secrets; copyright; and the database right. The protection afforded by these IPRs may be limited in certain circumstances and may also vary between jurisdictions. The existence of such IPRs and the implications they may have on the use of data can be dealt with by having in place appropriate licensing terms.

There is a common misconception that information can be protected by IPRs alone. As discussed earlier, while there may be some IPRs present in certain datasets, these provide limited protection. This is particularly relevant in the context of a data centre where multiple parties may be contributing and extracting potentially valuable datasets. Under English law, there is no property in information even if that information is confidential. Therefore, a claim against a third party for misuse of confidential information requires a pre-existing duty to keep the information confidential such as a contractual right. For greatest certainty, the data needs to be considered distinctly from intellectual property and protected under contract.

As the financial investor does not own the data by default, he needs to build and maintain data 'ownership' into the contract documents. This means being wary of contractual terms that may erode that data ownership. For example, combining datasets will typically increase their value so consideration must be given to the licensing of third-party data and whether there are limitations on its future use. An investor will want to avoid limitations in the contract documents on subsequent commercial uses or sub-licensing of combined datasets.

In the litigation section mentioned earlier, we refer to dispute risks that may arise such as an allegation that a third party's intellectual property is being infringed in some way by the data centre operator. Such an allegation could be directed to the data system or the data itself. Claims against the system could involve a copyright infringement in respect of the software in use or a patent infringement claim in respect of the system's hardware. An example of data related dispute would be a claim that copyrighted content stored on the data centre's servers is being accessed by unauthorised users and therefore infringing copyright and/or database rights. If that data is also confidential, then misuse of confidential information claims may also arise. The legal due diligence process can be used to identify any actual or threatened claims. The risk posed by any potential disputes that are identified can then be assessed prior to acquisition. The financial investor can also attempt to mitigate future losses through due diligence and by seeking appropriate warranty and indemnity provisions in the contract documents.

Data and Cyber/Regulatory Compliance

As with any centralised data storage facility, data centres are potential targets for cyber attacks. Given the high stakes – both in terms of liability and public relations – cyber security is key to attracting and reassuring customers.

Alongside this, an efficient and up-to-date compliance regime – including in respect of the processing of personal data – will go a long way to minimising the risk of potential liabilities arising from a regulatory breach.

Regulatory regimes must be identified and understood, and associated risk and compliance issues considered on a case-by-case basis. What regimes may come into play and how associated risks and compliance are managed will depend on numerous factors such as the identity of the parties to the arrangements, nature of the data and services, geographical location of the data centre/services, and potential regulator/state authority and action. First and foremost, the storage of personal data will need to be compliant with data protection laws. The territorial scope of GDPR covers not only the activities of an establishment located in the EU but also any processing activities relating to the offering of goods or services to data subjects in the EU. Where the customer is a provider of essential services (e.g. electricity, water, transport) it may be subject to specific regulations such as the United Kingdom's NIS Regulations, 2018 (which also cover certain digital services including online search engines and cloud computing services). In this case, the customer may need/seek to flow certain security obligations down to the data centre provider. Additionally, any industry-specific regulations, such as in the insurance or financial services sectors, should also be considered carefully.

Disputes

Data centre disputes may take many different forms. In terms of infrastructure disputes, these can relate not only to basic issues such as the workmanship of the bricklaying but also to the design of the mechanical and electrical systems that are needed to ensure the servers are able to operate without interruption. Seeking to place all design responsibility on a contractor who may not have the necessary expertise will cause problems, and we frequently see supply chain management issues arising. Operationally, problems may be encountered in the service provider achieving all SLAs and KPIs and transition disputes may arise in connection with delays and technical difficulties arising from the migration of data and platforms from one data centre to another, be they apps, distributed or mid-range technologies, mainframe, or otherwise.

In the intellectual property section mentioned earlier, we refer to the types of IP disputes that may arise. Namely, allegations that the data centre's operation infringes a third party's rights in either the data centre's systems or the data itself. However, an operator should also ensure that its intellectual property is identified, and a strategy put in place to protect it.

A dispute with a service provider can also negatively impact access to the data. In the recent case of *Mott MacDonald Ltd. v. Trant Engineering Ltd.* [2021] EWHC 754 (TCC), Mott MacDonald was contracted to provide design services including BIM development. Mott MacDonald believed it had carried out the services it was contracted to provide but then a dispute arose in respect of the work undertaken and, consequently, about payment. As a result, Mott MacDonald revoked the passwords which they previously provided to Trant and which had granted Trant access to the building information modelling database which Mott MacDonald had designed. Although the case did not involve a data centre, it is illustrative of the access issues that could arise.

One recent dispute concerned intellectual property and confidentiality. In 2011, BladeRoom Group Limited ('BladeRoom'), a UK-based business specialising in modular data centre design, entered into discussions with one of its competitors, Emerson Electric Co. ('Emerson') relating to a potential sale of BladeRoom to Emerson. Those discussions eventually fell through.

Around the same time, Facebook began planning a large data centre in Northern Sweden. Both BladeRoom and Emerson pitched data centre designs to Facebook. In 2013, Facebook selected Emerson's proposal and the two signed a design-build contract in 2014.

In 2015, BladeRoom sued Facebook, alleging the data centre's design copied BladeRoom technology. These designs were subsequently made publicly available through the Open Compute Project, a non-profit initiative led by Facebook, the stated aim of which is to facilitate the sharing of data centre product designs and best practices among industry players. BladeRoom subsequently added Emerson as a party to its claim and asserted various claims against Emerson and its subsidiaries.

The element of the dispute involving Facebook was settled in 2018. Later that year, the district court handed down judgment against Emerson Electric, ordering it to pay BladeRoom an amount of US$77.4m. Subsequent to that judgment, however, the Court of Appeals for the Ninth Circuit overturned the verdict and ordered a new trial.

While the key issue in that case revolved around whether or not a non-disclosure agreement remained in force, the case provides a salutary warning more generally, about the importance of intellectual property and the need to protect it.

A further dispute arose over the well-publicised data centre outage in May 2017 that forced British Airways ('BA') to ground hundreds of flights and reportedly left 75,000 passengers stranded. At the time, the data centre in question was reportedly operated by a subsidiary of CBRE, the US-based property services and investment business. The details are not public as BA and CBRE reached a commercial agreement with no admission of liability in respect of the outage, but this case does illustrate not only the financial implications of a data centre outage but also the reputational issues that may arise.

Conclusion

In conclusion, we can see that the legal issues relating to data centres cross many different disciplines and can give rise to many different concerns.

The earlier these issues can be identified and addressed, the better. In some cases, the solution will be capable of being designed into the process and dealt with upfront – whether in respect of planning, procurement, or infrastructure connectivity.

And as in all regulated industries, the ability to navigate through these issues can become its own differentiator in the market.

From a legal perspective, the services clients require are as varied as the issues to which data centres give rise and clients benefit from a holistic approach that is not artificially limited to any single aspect of the procurement or lifecycle of the asset.

11

Around the Corner, What Could Happen Next

Data centres enable people, communities, and businesses to function and thrive. They underpin our digital society and, increasingly, the way we live. Demand for data processing is expected to continue to grow and despite the trend of increasing server energy efficiency, the result has still been a net increase in energy consumption. Increasing smartphone users, IP-connected devices per household, use of IoT (Internet of Things) devices and artificial intelligence processing of datasets all increase demand.[1] Technological developments and economic drivers will continue to shape the data centre landscape.

Availability of IT services remains a key requirement – perhaps more so as our dependence on data processing continues to grow. Threats to availability are continuous; operators must stay alert to new ways in which they may be targeted. Increasingly, breaches of cyber security such as DNS attacks have caused high-profile outages.

It seems likely that the hyperscalers will continue to grow their market share with increased convergence into the cloud. This is in part due to their ability to achieve economies of scale and technology-focus (it is their core business) which delivers price competitiveness compared with self-delivery.

At the other end of the scale, demand for edge computing close to end users may grow decentralised computing where small facilities proliferate in urban environments. The shift towards home working has reduced demand for city-centre office space which makes alternative uses economically more attractive.

---

1 Ferreboeuf, H. (n.d.). *The shift project, think tank de la transition carbone 1 5 Novembre 2017 – IA, Numérique et Environnement Intelligence Artificielle, Numérique et Environnement.* [online] Available at: https://theshiftproject.org/wp-content/uploads/2017/11/note_danalyse_synthese_de_laudition_par_la_commission_villani-ia_the_shift_project.pdf [Accessed 23 Oct. 2022].

Data Centre Essentials: Design, Construction, and Operation of Data Centres for the Non-expert, First Edition. Vincent Fogarty and Sophia Flucker.
© 2023 John Wiley & Sons Ltd. Published 2023 by John Wiley & Sons Ltd.

Data centres and data transmission networks are responsible for nearly 1% of energy-related greenhouse gas emissions.[2] Developments in the energy sector, such as advances in battery technology and hydrogen fuel cells should increase the use of renewable energy for data centres and electricity grids in general. Policy-makers are becoming increasingly aware of data centre resource consumption and are acting through regulation, even suggesting that streaming sites contribute to investment in internet infrastructure.[3] However well-intentioned, policy can be a blunt instrument which is unable to keep up with the pace of change. Technology that does not respect borders and rules in one region may simply displace the problem elsewhere. Legislators need to balance the requirement to support the digital infrastructure that data centres provide with their environmental impact and demands from other resource users.

Increased regulation will increase awareness and reporting of the environmental impact of data centres, including embodied impacts. Public emissions reporting by large corporations is helping to drive change. There is interest in moving away from burning diesel in standby generation to reduce Scope 1 emissions. Microsoft aims to eliminate the use of diesel by 2030[4] and has piloted hydrogen fuel cells as an alternative.[5]

Members of the public and users of data remain relatively detached from the impact of their consumption of data and its associated environmental footprint; after paying for a data subscription, most digital services remain free. There is little transparency which allows consumers to compare the relative environmental footprint of different websites. Not only would it be difficult and unpopular to curtail end user consumption with usage restrictions but it is more effective to bake sustainability into the design process, starting with software development.[6] It is hoped that this will change through increasing environmental awareness from all stakeholders. The criticality of addressing climate change is becoming more acute, and the public as well as regulatory pressure may help drive the

2 IEA (n.d.). *Data centres & networks – fuels & technologies.* [online] Available at: https://www.iea.org/fuels-and-technologies/data-centres-networks.

3 Data Center Knowledge | News and Analysis for the Data Center Industry (2022). *Key EU policymakers want Netflix to pay more for infrastructure.* [online] Available at: https://www.datacenterknowledge.com/companies/key-eu-policymakers-want-netflix-pay-more-infrastructure [Accessed 23 Oct. 2022].

4 Microsoft on the Issues (2020). *Progress on our goal to be carbon negative by 2030.* [online] Available at: https://blogs.microsoft.com/on-the-issues/2020/07/21/carbon-negative-transform-to-net-zero/.

5 Innovation Stories (2020). *Microsoft tests hydrogen fuel cells for backup power at datacenters.* [online] Available at: https://news.microsoft.com/innovation-stories/hydrogen-datacenters/.

6 Principles.Green (n.d.). *Principles of green software engineering.* [online] Available at: https://principles.green/.

industry to improve its sustainable performance not just in terms of environmental impact but also social impacts such as human rights. The business case for data centre sustainability continues to grow.[7]

Liquid cooling of IT equipment is growing in market share as chip densities increase. This represents the paradigm shift in how cooling is delivered which not only has the potential to reduce environmental impact but also comes with a number of challenges. The design, installation, commissioning, and operation (including failure modes) of liquid-cooled systems is different to that of air cooling and will be a learning process for all those involved and force the world of IT hardware thermal design to become closer to that of cooling system design. Due to the improved heat transfer properties of liquid compared with air, cooling takes place at higher temperatures. There are different types of liquid cooling, often with a dielectric fluid which rejects heat via a heat exchanger to a water system, using a cold plate in contact with the components or immersion cooling. This allows designs where heat can be rejected in any climate without the use of refrigeration, resulting in energy as well as material savings and assisting with adaption to increasing ambient temperatures. However, ASHRAE warns that increased heat densities will require lower temperatures, reversing cooling efficiency trends.[8] Liquid cooling can lend itself to capturing higher-grade heat which extends the opportunities for heat recovery/waste-heat reuse.

Quantum computing uses quantum mechanics to perform computations and for certain complex applications (e.g. cryptography, machine learning, and drug development), can potentially solve computational problems substantially faster than classical computers. However, there are technical challenges in building a quantum computer at scale including the availability of parts and stability, which is in part achieved by controlling the processor temperature to about a hundredth of a degree above absolute zero. Quantum computing makes use of superconductors – where electrons move through processors at ultra-low temperatures without resistance. It requires much less energy than classical supercomputers. However, most of the hardware is made up of cooling systems to maintain the required ultra-cold operating temperature. Should it become possible to overcome these challenges and allow large-scale exploitation of this technology, this will fundamentally change computation as we know it and thus the data centre landscape.

---

7 Tozer, R., Flucker, S. and Whitehead, B. (2016). *The case for sustainability in data centers.* ASHRAE Conference Proceeding OR-16-008.

8 White Paper Developed by ASHRAE Technical Committee 9.9, Mission Critical Facilities, Data Centers, Technology Spaces, and Electronic Equipment (2021). *Emergence and Expansion of Liquid Cooling in Mainstream Data Centers.* ASHRAE.

If room-temperature superconductors become viable, this could lead to a paradigm shift in power distribution and generation, which again would result in huge changes in data centres.

Perhaps another, as yet unknown, technology will disrupt the status quo. It is important for those working in data centres to continue looking to the future in anticipation of the changes to come. Each new facility is different to the one before; it is essential that the industry keeps adapting to the challenges that it faces. Learning lessons from past experiences and understanding the changes to come can help deliver flexible, efficient data centres with reduced operating costs and improved sustainability, which are ready to meet future requirements. We hope that this book will help prepare you.

Glossary

Adiabatic A process where the air temperature is decreased and humidity is increased by the evaporation of water

Aisle A corridor between rows of racks

Artificial intelligence Is software or machines that can complete tasks and solve problems by thinking like a human being

Big Tech Big Tech are often referred to as major foreign cloud service providers that dominate market share through economies of scale and vertical integration, namely Alphabet (Google), Amazon, Alibaba, and Microsoft. It further refers to digital product and service companies which have a significant market share in their respective category, for example, Tencent or Facebook.

Bitcoin A digital currency that acts as a form of payment for electronic transactions without passing through traditional financial intermediaries like banks or clearing houses.

Blockchain A mathematical structure for storing digital transactions or data in an immutable, distributed, decentralised digital ledger consisting of blocks that are linked via cryptographic signature that is almost impossible to fake, hack, or disrupt.

Buffer vessel Tank included in the chilled water system, which acts as a thermal store

Busbar The solid metal bar that carries current (alternative to cable)

Cabinet The metal rack where IT hardware is housed

Changeover device Electrical switching device which allows an alternate electrical source to supply a connected load when the primary source fails

Chiller Refrigeration machine which cools chilled water

Closed transition When the transfer from the generator back to the mains supply is no break

Data Centre Essentials: Design, Construction, and Operation of Data Centres for the Non-expert, First Edition. Vincent Fogarty and Sophia Flucker.
© 2023 John Wiley & Sons Ltd. Published 2023 by John Wiley & Sons Ltd.

Cloud Computing The paradigm of using digital commodity resources (see Cloud infrastructure) to perform computations – e.g. to run a software application, a function, or any form of software-driven computation.

This paradigm is different from traditional computing paradigms in which the server and equipment are dedicated or at least known to the user or the software application. In Cloud computing, digital resources can come from anywhere (as they are commodities) – e.g. CPUs can be located in one server, and storage can be located in a different server in a different data centre. In this paradigm, it is neither possible nor desirable to know the physical server or location, as it would break the commodity principle of the resource.

Cloud Infrastructure The paradigm on the separation of digital resources (computing power, networking, and storage) from the physical infrastructure (servers, data centre, and energy) and providing those resources as a consumable commodity to software applications.

Cluster A Cluster is a collection of Data Centres

Colo(cation) A facility where IT space is rented out

Commissioning The process of setting up and testing equipment and systems

Containment A physical barrier between hot and cold air streams. Also used to describe cable trays.

Data The means by which software applications via application programming interfaces (APIs) or user interfaces (front-ends and clients) communicate with each other, receive user inputs, and store inputs and outputs.

Digital Economy For the term 'digital economy', we use the definition of the Eurofound (Eurowork, 'Digital economy', 2017):

When applied to social systems or organisations, digitalisation refers in a broad way to the transformation brought about by the widespread adoption of digital technologies (robotics, machine learning, sensors, virtual reality, and so on). When applied to specific pieces of information, digitisation refers to the process of converting them into a digital form that can be processed by a computer (or vice versa): it can be seen as a driver of digitalisation and encompasses actions performed through technologies such as sensors, 3D printing or augmented reality, and virtual reality.

Digital Infrastructure The total physical and software-based infrastructure necessary to deliver digital goods, products, and services. This includes data centres, fibre infrastructure, server hardware, personnel, IT virtualisation and infrastructure software, operating systems, etc.

Digital Product A digital product or service is what a company or a consumer might buy. It is always made up of software or a combination of software and hardware and, therefore, deals with data, which represents its inputs (by the users), which it transforms into outputs. The quality of a digital product is often measured by how well it does this conversion of inputs to outputs, how those outputs are visualised, how easy it is to input something, how fast it does the conversion, and with what rate of error.

Digital resource primitive Digital resource primitives are defined as the low-level resources required for digital products and services to operate. They can be seen as the fuel that powers digital software applications. The digital resource primitives are:

Computation (CPU cycles, GPU cycles, and operations per second)
Memory
Storage
Network capacity
Digital resource primitives are often accessed through Cloud infrastructure, also commonly referred to as Infrastructure-as-Service.

Downtime When equipment/system is not available

Dual-corded Equipment with two power supplies

Edge Edge computing is an emerging computing paradigm which refers to a range of networks and devices at or near the user

Economiser cooling See Free cooling

Embodied impact Impact related to the production and disposal of a product or process rather than the impact during its use phase

Enterprise An enterprise data centre is a facility that an organisation operates to support its data processing and storage needs

Environmental envelope Range of air temperatures and humidities

Fibre Refers to the technology that transmits information as light pulses along a glass or plastic fibre.

Free cooling Compressor-free cooling, which does not require refrigeration

Handover When the project is completed and the project is handed over to the client

Heat pump A device which heats using a refrigeration cycle (opposite to cooling)

Heat recovery Collection of heat extracted from cooling IT hardware to be used for heating elsewhere

Hyperscale A computing architecture which allows scaling with the demand of computing, memory, networking, and storage

Internet of Things It is a network of physical devices, appliances, and other items integrated with electronics, software, sensors, and more that digitally collects and exchanges data about the physical world

Latency The time between a command issuing and a server responding in return trip time

Legacy facility A facility which is a few years old using older design/technology

Legionella The bacterium which causes Legionnaire's disease

Load bank Test equipment used to simulate an electrical/thermal load

Modular Build A process where a building is constructed offsite using controlled plant conditions before being transported and assembled at a final location

Moore's law Every two years, computational speed and capability double due to increasing number of transistors on a microchip

Open transition When the transfer from the generator back to the mains supply has a break

Outage Failure resulting in a service interruption

Pseudomonas Bacterium, which may be found in water systems. Production of biofilm may cause cooling performance as well as hygiene risks.

Psychrometric chart Graphical method for representing temperature and humidity of air

Procurement Procurement in construction is the process of securing all the goods and services needed to bring a construction project to completion

Rack See cabinet

Redundancy Used to describe spare capacity equipment or systems

Refresh rate How often old IT hardware is replaced with new

Remote hands Service where customers pay for on-site technicians to undertake IT management and maintenance tasks

Soft landing Building a delivery process which aims to improve the transition from project to operation, resulting in a high-performing facility

Switchgear Large electrical distribution panel, which includes switches

Switching When electrical breakers are open/closed

Switchroom Plant room housing electrical switchboards

Technical submission A document which includes details of equipment proposed for use. The designer usually approves that this is in conformity with the specification

Thermal inertia The ability of a space/system to absorb heat, slowing down temperature increase

Thermal runaway When data hall temperatures increase due to power failure

Tier level Used to describe design reliability of critical infrastructure (1 lowest, 4 highest)

Uptime When equipment/system is available

Virtualisation Creation of a virtual rather than a physical version of the server, storage or network resources.

White space Also known as the data hall, where the IT hardware is housed within the facility.

Index

Data Centre Essentials: Design, Construction, and Operation of Data Centres for the Non-expert, First Edition. Vincent Fogarty and Sophia Flucker.
© 2023 John Wiley & Sons Ltd. Published 2023 by John Wiley & Sons Ltd.